Key Questions in Urban Pest Management:
A Study and Revision Guide

FSC
www.fsc.org
MIX
Paper | Supporting
responsible forestry
FSC® C022174

Key Questions in Urban Pest Management: A Study and Revision Guide

Partho Dhang
Independent Consultant, Philippines

Philip Koehler
Emeritus Professor, Entomology & Nematology Department, University of Florida, USA

Roberto Pereira
Extension Professor, Urban Entomology Laboratory, Entomology & Nematology Department, University of Florida, USA

Daniel D. Dye II
Photographer, Florida, USA

CABI is a trading name of CAB International

CABI
Nosworthy Way
Wallingford
Oxfordshire OX10 8DE
UK

Tel: +44 (0)1491 832111
E-mail: info@cabi.org
Website: www.cabi.org

CABI
WeWork
One Lincoln St
24th Floor
Boston, MA 02111
USA

Tel: +1 (617)682-9015
E-mail: cabi-nao@cabi.org

The views expressed in this publication are those of the author(s) and do not necessarily represent those of, and should not be attributed to, CAB International (CABI). CAB International and, where different, the copyright owner shall not be liable for technical or other errors or omissions contained herein. The information is supplied without obligation and on the understanding that any person who acts upon it, or otherwise changes their position in reliance thereon, does so entirely at their own risk. Information supplied is neither intended nor implied to be a substitute for professional advice. The reader/user accepts all risks and responsibility for losses, damages, costs and other consequences resulting directly or indirectly from using this information.

CABI's Terms and Conditions, including its full disclaimer, may be found at https://www.cabi.org/terms-and-conditions/.

A catalogue record for this book is available from the British Library, London, UK.

Library of Congress Control Number: 2022935435

References to Internet websites (URLs) were accurate at the time of writing.

ISBN-13: 9781800620155 (paperback)
         9781800620162 (ePDF)
         9781800620179 (ePub)

DOI: 10.1079/9781800620179.0000

Commissioning Editor: Ward Cooper
Editorial Assistant: Lauren Davies
Production Editor: Tim Kapp

Typeset by SPI, Pondicherry, India
Printed and bound in the UK by Severn, Gloucester

Contents

About the Authors

Partho Dhang received his B.Sc., M.Sc. and Ph.D. in zoology from University of Madras, Chennai India. For his doctoral thesis, he worked on insect physiology and natural products. He joined a private organization, SPIC Science Foundation in India, as a scientist and worked on development of various bio-rational and crop protection products including plant-based bio-pesticides, insect pheromones and microbial larvicides. Most of the products were commercialized during his tenure. He left India in the year 1998, to work for a number of companies on short stints, mainly focused on urban entomology. Prominent among his jobs was a Singapore government funded project on emerging technologies on control of urban pests. Then in 2005 he moved to the Philippines to set up his own consultancy work. His close association with the pest control industry, covering works such as research and development, training and business development has allowed him to edit and write a number of books, all published by CABI. He is also a prolific speaker at international conferences across the world and has spoken at National Pest Management Association (NPMA), International Conference on Urban Pest (ICUP), International Research Group in Wood Protection (IRGWP) and major conferences all over the world. He is a regular contributor of articles to various international magazines. Presently he serves on the panel of judges for the award of doctoral degree for two universities in India in the subject of economic entomology. He is also a technical consultant for International Pest Control magazine and an Executive Committee member of Pacific Rim Termite Research Group (PRTRG). Presently he resides in the Philippines and India. E-mail: partho@urbanentomology.com

Philip G. Koehler is an urban entomologist with a B.A. in Biology from Catawba College and a Ph.D. in Entomology from Cornell University. He then became a Lieutenant, Medical Entomologist, in the Medical Service Corp. with the U.S. Navy. After 3 years of service, he joined the University of Florida as Professor and Extension Entomologist and remained in that position for 45 years. He is now Professor Emeritus of Urban Entomology at the University of Florida, a fellow of the Entomological Society of America, and a fellow of the National Academy of Inventors. He is an honorary member of the Florida Pest Management Association and honorary member of the Certified Pest Control Operators of Florida. His honors include the National Pest Management Association's Leadership Award, the University of Florida's Distinguished Faculty Award, University of Florida's Academy of Teaching Excellence, University of Florida Professorial Excellence Award, 2 USDA Distinguished Service Awards, and the Florida Pest Management Association's Outstanding Service Award. He has published over 200 scientific publications, given over 3000 presentations to the pest control industry, and is well known for his research in urban entomology and structural pest control, including research on bed bugs, flies, mosquitoes, cockroaches, termites and ants. E-mail: pgk@ufl.edu

Roberto M. Pereira is an entomologist with a B.Sc. in Agriculture from the University of São Paulo (Brazil), M.Sc. in Entomology from Cornell University (USA) and a PhD from the University of Florida (USA). Early in his career, Roberto studied microbial control of insects including sugarcane and pasture pests, fire ants and other urban pests. Roberto is an Extension Professor at the Urban Entomology Laboratory, University of Florida. He conducts research, education and extension in several areas of urban entomology, including ants, cockroaches, termites, bed bugs, flies, mosquitoes and other pests. Roberto has authored or coauthored more than 130 scientific publications and book chapters, numerous non-refereed and trade journal publications, and more than 20 US and international patents in pest control. He is the Managing Editor and contributor for PestPro Magazine, a bimonthly magazine dedicated to the urban pest management industry, freely available at PestProMagazine.com. Roberto is a member of the Entomological Society of America, the Society for Invertebrate Pathology, the Entomological Society of Brazil, and the Florida Entomological Society. E-mail: rpereira@ufl.edu

Daniel D. Dye II is an accomplished photographer. His insect and herpetofauna photos have been published in several books, magazines and articles. Although his main expertise is in urban entomology, he has a keen interest in other insects, spiders and creepy critters found in wild places. He became interested in the small creatures of Earth at an early age while chasing rabbits and catching frogs in his grandfather's field in Germany. His grandfather called him "der Froschjunge," which translates to "the frog boy" in English. Daniel would often bring home jars filled with insects, much to his mother's dismay. In 1977, Daniel was introduced to pest control by ARAB Termite & Pest Control in Tampa, Florida. There he learned about wood-destroying organisms, general household pests, and pests of turf and ornamentals. Daniel is an Associate Certified Entomologist and holds four Florida State

Certifications: Termite and Other Wood destroying Organisms – Fumigation – Lawn and Ornamentals – General Household Pests and Rodent Control. In 1995, Daniel took a position with Florida Pest Control and Chemical Co. In 2002, he became the Training Coordinator based at the corporate office in Gainesville, Florida. During his time at Florida Pest Control, he served on the Board of Directors for the Florida Pest Management Association. Daniel retired from Florida Pest Control in December of 2017. To keep his mind sharp, he is a frequent speaker and consultant for pest management companies and associations. When not photographing arthropods, Daniel and his wife, Yvonne, search for herpetofauna to photograph. E-mail: dydmdcloud@icloud.com

Preface

The subject of urban pests and their management could be considered as old as human civilization. It has passed through an era where it was dealt with magic and religion. In the seventeenth century, science and experimentation opened up new understanding of pests, such as their physiology, behaviour and function in ecosystems. Up until the mid-19th century their control depended mostly on plant derived products, along with inorganic compounds. The next era of pest control started with the discovery of DDT (a synthetic organochlorine compound) which proved to have controlling power over a broad variety of pests. DDT was later banned along with others dues to environmental concerns. Today pest management continue to depend on synthetic organic chemicals, but with compounds proven to have lesser environmental and health impact.

The subject, in spite of enormous developments in control methods, has remained relevant to humans. More reports of pest related diseases, injuries and damages have been gathered in recent times than in the historical past. However, many aspects of pest's presence in the vicinity are human induced. Humans transported insects like the German cockroach and termites across the world and human behaviour of collecting water, creating clutter and generating waste are reasons for mosquito, fly and rat menace. All of these make the subject of urban pest management complex and in turn make the work of pest managers critical.

This book is designed as a general revision guide with a little over 500 key questions and answers, in our effort to make knowledge on urban pest management handy to review, and available in a trainable form. It was indeed a challenge to limit our choice of questions in a subject which is vast, interdisciplinary and modernizing at a pace never seen before.

The book is suitable for beginners and graduate students, including pest management technicians, field workers, affiliated and support staff of pest management companies. Also, trainers can use this book in conducting their education exercises, including laying out tests.

It is our sincere hope that this book serves its purpose well.

Partho Dhang
Philip Koehler
Roberto Pereira
Daniel Dye II

Acknowledgements

We wish to thank CABI, specifically Ward Cooper, for suggesting the project. It was indeed timely, as the pandemic continues to disrupt methods of learning, and any type of alternative tool is welcomed.

We also would like to thank the entire production team at CABI for their support, by working under challenging times.

Books Referred to and for Further Reading

Dhang, P. (ed.) (2011) *Urban Pest Management: An Environmental Perspective*. CAB International, Wallingford, UK.

Dhang, P. (ed.) (2014) *Urban Insect Pests: Sustainable Management Practices*. CAB International, Wallingford, UK.

Dhang, P. (ed.) (2016) *Climate Change Impacts on Urban Pests*. CAB International, Wallingford, UK.

Dhang, P. (2018) *Urban Pest Control - A Practitioner's Guide*. CAB International, Wallingford, UK.

Doggett, S.L., Miller, D.M. and Lee, C.-Y. (eds) (2018) *Advances in the Biology and Management of Modern Bed Bugs*. Wiley-Blackwell, Hoboken, New Jersey.

Wang, C., Lee, C.-Y. and Rust, M. (eds) (2021) *Biology and Management of the German Cockroach*. CSIRO and CAB International, Wallingford, UK.

Introduction to Urban Pest Management

1

As little as 1% of the earth's total land mass has been transformed into urban centers called cities, and these cities astoundingly carry 50% of the world's population. It is estimated that by 2050, almost 70% of all people will live in cities. Consequent to this development, urbanization is now growing vertically into high rise buildings as well as horizontally into the natural habitats surrounding cities. Such development brings concentrations of people and activities, producing enormous waste. This waste creates countless man-made niches and micro-habitats which together make urban areas susceptible to pest invasion and long-term harborage.

Urban centers are extremely well suited for a group of invertebrate and vertebrate organisms that have associated their lives with humans and their activities. These organisms cause pain, annoyance, emotional distress, disability and loss to humans as a result of bites, stings and physical reactions, in addition to a plethora of diseases and significant damage to buildings. All of these organisms collectively are called urban pests. Urban pests include household, structural and public pests and this distinguishes them from agricultural and forest pests.

Need for Pest Management

Pest control is becoming a necessity for humans. Apart from diseases, the sight of pests triggers various types of negative behavior: anger, disgust and, almost inevitably, the use of a toxic chemical spray. This human behavior has made pest management an easily tradeable business. However, the degree of the trade is dependent on the nature of the service the practitioner is offering and the environment where the service is required. An occasional trail of ants in the home may be a mere nuisance; in contrast, a single ant in a hospital

© Partho Dhang, Philip Koehler, Roberto Pereira and Daniel D. Dye II 2022. *Key Questions in Urban Pest Management: A Study and Revision Guide* (P. Dhang et al.)
DOI: 10.1079/9781800620179.0001

can have serious consequences. The tolerance to pest infestations varies from situation to situation. In comparison to a home or shopping center, institutional kitchens, healthcare facilities and critical manufacturing complexes demand detailed and careful design and planning to exclude pests.

Types of Urban Pests

Urban pests are categorized in various ways, and one method is by the nature of their interaction with humans. This is a useful and convenient approach, as by far the ideal definition of a pest is, "any organism that appears in a place where it is unwelcome to humans". This definition further clarifies that pest status does not adhere to any taxonomic line, such as orders or families, nor a location.

Following are the most notable pest categories:

Pests associated with human blood

Insects representing half a dozen orders use humans as a source of food. Direct blood feeders such as mosquitoes and bed bugs rank top in the group of insects causing intentional injury. The next group are the ecto-parasites such as lice, fleas and ticks which cause injury and diseases. Ticks are known to transmit Lyme disease, tick-borne encephalitis and also tick paralysis. Similarly, fleas are associated with plague, and lice with typhus.

Pests inflicting injury

Envenomation by bees, ants and wasps is another source of injury to humans which at times could be fatal. Insect venom is considered a leading cause of human mortality through direct injury by arthropods.

Pests associated with allergens, contamination and phobia

A number of pests have gained significance by becoming known as sources of allergens, food contamination and entomophobia.

Pests of stored products

These are a relatively inconspicuous group which humans encounter in stored products. Stored items such as food, clothing, furnishings, artifacts and books are continuously attacked by these groups of pests. They include insects, rodents and birds.

Pests of buildings and structures

Insect pests of structures and buildings have made themselves notable by using parts of human dwellings as food and shelter. Termites and powder-post beetles regard wood used in construction as well as furnishing as potential food. Ants living in soil too have become a major structural pest in recent times. Rodents, bats and birds are also significant pests of structures as they cause damage, defacement and disease.

The History of Urban Pest Management

The history of human interaction with pests goes back to the beginning of civilization. Both insects and rodents have been responsible for disease and death in humans, and ever since efforts to minimize their interaction with humans have grown to be an important public and private undertaking. Homes are sealed, sprayed and kept clean; bodies are bathed, clothing washed, food cooked and garbage disposed of in order to maintain distance from these pests.

The earliest record of urban pest management as a practice can be identified from reported attempts made by Romans to drain marshes to control malaria. In 18th and 19th century Europe, rat trapping had become specialized work. Soon, bed bug elimination and control of timber pests also became a type of professional work. It is reported that exterminators from parts of Europe carrying knowledge of such pests as rats, bed bugs and timber pests emigrated to the United States during this period, laying the foundations for the American urban pest control industry.

It was the development of DDT in the 20th century which ushered in the modern era of pest management. It belonged to the second generation of pesticides, synthetic compounds, which followed the first generation of inorganic compounds, such as heavy metals and plant botanicals. DDT was first utilized in suppressing a typhus epidemic in Italy in 1943–1944 and then used in reducing deaths from malaria and other insect borne diseases during war in the Pacific islands. Its efficacy and cheap production costs led to the development of a variety of chlorinated hydrocarbons, and later organophosphates. These chemicals became part of most pest management programs, until the publication of Rachel Carson's book "Silent Spring". Both government and the public took serious note of the fact that the indiscriminate use of pesticides was causing harm to the environment,

3

specifically non-target organisms. This realization was soon followed by the formation of regulatory bodies and enactment of laws all around the world which looked into the judicious use of pesticides. It also led to the development of the third and fourth generations of pesticides and the concept of integrated pest management (IPM). The third generation includes all biochemicals which influence insect development, such as juvenile hormone (JH) and chitin synthesis inhibitor (CSI). The fourth generation includes chemicals which modify insect behaviour, such as insect pheromones and antifeedant compounds.

Urban pest management is not restricted to chemicals and chemical control methods. There are a number of tools such as exclusion devices, engineering concepts, pest-proofing materials, repellents, monitors and traps which are equally useful in managing pests. The best management is achieved when all of these are combined and used in an integrated pest management (IPM) program.

Today, the urban pest management industry faces challenges from many factors such as increasing insecticide resistance in pests, surging bed bug numbers, new research regarding the toxicity of chemicals, stricter government regulations, environmental concerns amongst the public and climate change.

Urban pests are also the cause of lawsuits around the world. It is somewhat unavoidable, as these pests can enter and establish themselves in the unlikeliest of places. Disputes of this nature are resolved by experts through understanding the correct behavior and biology of the pests.

Key Questions

1.1   The most appropriate definition of a pest is

> a. an organism that causes injury or harm to humans
>
> b. an organism that appears in a place unwelcome to humans
>
> c. an organism that enters a home
>
> d. an organism that belongs to Arthropoda

1.2   Certain organisms are categorized as urban pests as they are

> a. present in an urban setting
>
> b. associated directly with man and man-made objects
>
> c. not in the list of agricultural pests
>
> d. associated with human since ancient times

1.3 **Types of organisms that fall under the urban/household pest category are**

a. only invertebrates

b. only arthropods

c. both invertebrates and vertebrates

d. microbes

1.4 **Pests are primarily attracted to structures by human activities, such as their behaviour and habits.**

a. True

b. False

1.5 **Some of the world's most intractable health problems are arthropod-borne diseases.**

a. True

b. False

1.6 **The best way to deal with delusory parasitosis is**

a. making sure that the infestation is indeed imaginary

b. treat the non-existent pest problem

c. refer to a psychologist or a psychiatrist

d. shift the patient to a new location

1.7 **Match the common reaction of humans to the respective pests.**

a. Cause allergy, annoyance

b. Emotional distress and damage

c. Disease and sickness

d. Phobia

1. Mosquito

2. Bed bug

3. Spider

4. Termite

1.8 **Choose the non-stinging and non-biting pest which is known to spread allergens.**

a. Dust mite

b. Housefly

c. Cockroach

d. Silverfish

1.9 **Allergic reactions from non-biting and non-stinging insects generally manifest as**

a. sneeze, cough

b. runny or stuffy nose

c. itchy eyes, nose, mouth or throat

d. muscular pain

1.10 **Envenomation is a major form of injury caused by arthropods which comes from**

a. bites

b. stings

c. accidental consumption

d. urticating hairs

1.11 **Match the following human disease with the pest it is associated with.**

a. Plague

b. Typhus

c. Malaria

d. Lyme disease

e. Sleeping sickness

f. Chagas disease

1. Ticks

2. Fleas

3. Triatome bugs

4. Anopheles mosquito

5. Lice

6. Tsetse fly

1.12   Certain species of mosquito are called "vector" as they are capable of

a. carrying germs on their bodies

b. harboring and transmitting disease-causing pathogens

c. inflicting a bite

d. causing fatality

1.13   When inspecting a report of an insect bite/sting, the following is a must.

a. Identification of the bite location on the body

b. Inspection of the site where the bite was inflicted

c. Consulting an expert

d. Sealing the suspected area

1.14   What is a sustainable way to limit pests in a structure?

a. Periodic use of pesticide

b. Keeping the entry points closed

c. Removal of all forms of food and water

d. Engaging a regular pest controller

1.15   Determining a "pest threshold" is important as it helps

a. determine the pest tolerance of the customer

b. prevents unwanted use of insecticides

c. eliminates the use of preventive spraying

d. makes the pest control job expensive

1.16   Match the following.

a. Termiticide

b. Molluscicide

    c. Acaricide

    d. Rodenticide

    1. Snails

    2. Termites

    3. Rats

    4. Mites

1.17   Which of the following belong to a first generation pesticide?

    a. Mineral oil

    b. Lindane

    c. Tobacco

    d. Pyrethroid

1.18   Juvenile hormone and its analogs are categorized as a

    a. third generation pesticide

    b. second generation pesticide

    c. fourth generation pesticide

    d. natural product pesticide

1.19   Organophosphates, carbamates and pyrethroids are part of a

    a. fourth generation pesticide

    b. second generation pesticide

    c. synthetic organic pesticide

    d. inorganic pesticide

1.20   Chitin synthesis inhibitors (CSI) are chemicals that

    a. disturb cuticle formation causing abortive molting and hatching defects

    b. have some have effect on fungi as a fungicide

    c. block a catalytic step leading to chitin synthesis

    d. belong to the category of third generation pesticides

    e. all of the above

1.21 **Chemicals like DDT and organochlorines were restricted and banned because**

   a. they caused fatality in humans

   b. insects showed resistance to them

   c. they caused environmental damage

   d. of the impact of the book "Silent Spring"

1.22 **Which of the following apply to insect pheromones?**

   a. They disrupt mating

   b. They attract and trap

   c. They generate a response on one sex only

   d. They are a fourth generation pesticide

1.23 **The first insect pheromone isolated and characterized was from a**

   a. silk worm

   b. house fly

   c. gypsy moth

   d. honey bee

1.24 **Spraying of chemicals is considered a mandatory job for technicians controlling structural pests.**

   a. True

   b. False

1.25 **What does IRAC stand for?**

   a. Insecticide Rotation Action Committee

   b. Insecticide Resistance Action Committee

   c. International Resistance Action Committee

1.26 **Match the following.**

   a. WHO

   b. FDA

   c. SDS

   d. GHS

1. Documents containing chemical hazard information

2. A system of classification and labelling of chemicals

3. Body responsible for international public health

4. Agency promoting public health through control and supervision of food and drugs

1.27   Resistance to an insecticide occurs in pests due to

a. repeated use of the same pesticide

b. spraying pesticide over the label dosage

c. changing climate

d. failing to spray regularly

1.28   The best way to overcome insecticide resistance is to

a. rotate chemicals with different modes of action

b. use multiple control methods at the same time

c. always spray according to label directions

d. monitor resistance

e. all of the above

1.29   The definition of "green" pest management differs between practitioners, but generally it refers to

a. using only natural products

b. safety and responsibility

c. less use of pesticide

d. use of alternatives to pesticide if available

1.30   Indoor application of pesticides, which are regulated by a complex risk assessment before and after they are put on the market, does not pose a high level of risk if the application of the product takes place according to label directions and with proper precautions.

a. True

b. False

2 Pest Identification

2.1 The mosquito shown in the photo is a(n)

    a. culex mosquito (*Culex quinquefasciatus*)

    b. yellow fever mosquito (*Aedes aegypti*)

    c. *Psorophora ferox*

    d. *Ochlerotatus atlanticus*

© Partho Dhang, Philip Koehler, Roberto Pereira and Daniel D. Dye II 2022. *Key Questions in Urban Pest Management: A Study and Revision Guide* (P. Dhang *et al.*)
DOI: 10.1079/9781800620179.0002

2.2   The mosquito shown in the photo is a(n)

a. asian tiger mosquito (*Aedes albopictus*)

b. yellow fever mosquito (*Aedes aegypti*)

c. *Psorophora ferox*

d. *Ochlerotatus atlanticus*

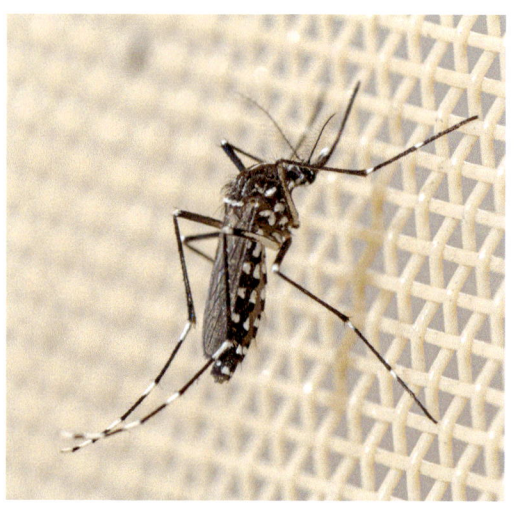

2.3   The mosquito shown in the photo is a(n)

a. asian tiger mosquito (*Aedes albopictus*)

b. yellow fever mosquito (*Aedes aegypti*)

c. *Psorophora ferox*

d. *Ochlerotatus atlanticus*

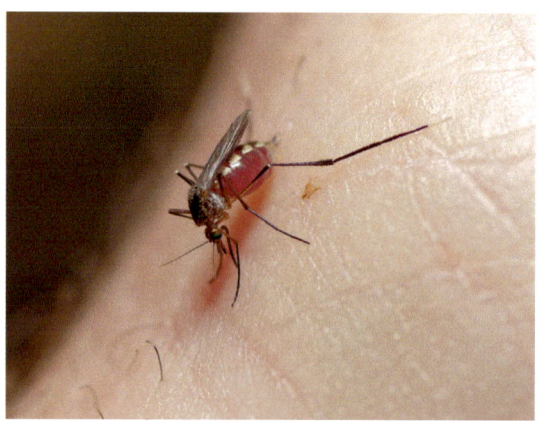

2.4   Match the images with the species

a. *Aedes* species

b. *Culex* species

c. *Anopheles* species

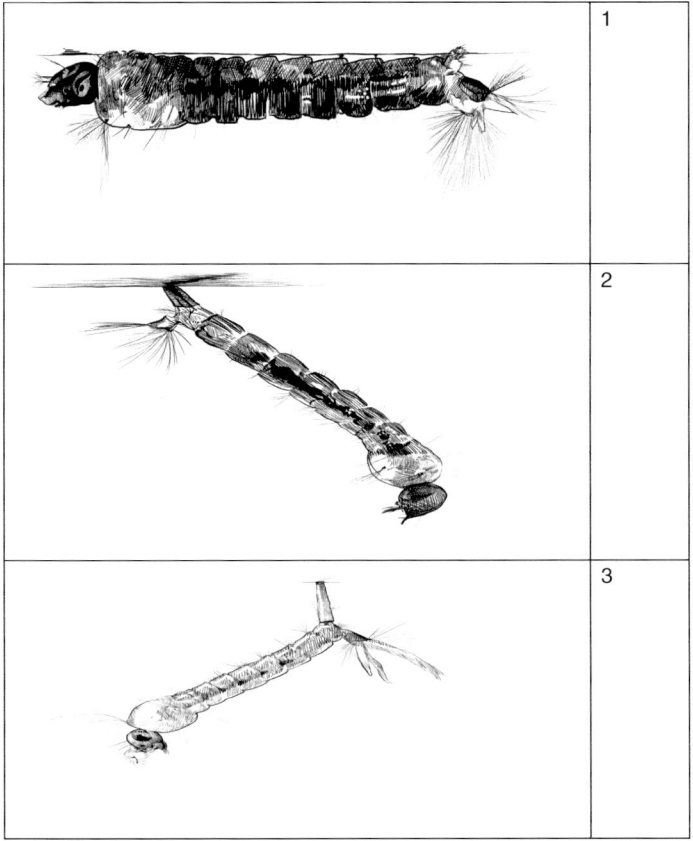

2.5   The *Cimex sp.* shown in the photo is a

a. common bed bug (*Cimex lectularius*)

b. tropical bed bug (*Cimex hemipterus*)

c. bat bug (*Cimex adjunctus*)

2.6 The fly shown in the photo is a

a. bottle fly (*Calliphoridae*)

b. fruit fly (*Drosophila*)

c. house fly (*Musca domestica*)

d. moth (drain) fly (*Clogmia*)

2.7   The fly shown in the photo is a

    a. bottle fly (*Calliphoridae*)

    b. fruit fly (*Drosophila*)

    c. house fly (*Musca domestica*)

    d. moth (drain) fly (*Clogmia*)

2.8   The fly shown in the photo is a

    a. bottle fly (*Calliphoridae*)

    b. fruit fly (*Drosophila*)

    c. house fly (*Musca domestica*)

    d. moth (drain) fly (*Clogmia*)

2.9   The cockroach shown in the photo is a(n)

    a. American cockroach (*Periplaneta Americana*) female

    b. brown-banded cockroach (*Supella longipalpa*) female

    c. German cockroach (*Blattella germanica*) female

    d. oriental cockroach (*Blatta orientalis*) female

2.10   The cockroach shown in the photo is a female

    a. brown-banded cockroach (*Supella longipalpa*)

    b. German cockroach (*Blattella germanica*)

    c. smokybrown cockroach (*Periplaneta fuliginosa*)

    d. Dubia roach (*Blaptica dubia*)

2.11   The cockroach shown in the photo is a(n)

    a. American cockroach (*Periplaneta Americana*) male

    b. brown-banded cockroach (*Supella longipalpa*) male

    c. German cockroach (*Blattella germanica*) male

    d. oriental cockroach (*Blatta orientalis*) male

2.12   The cockroach shown in the photo is a(n)

    a. American cockroach (*Periplaneta Americana*) female

    b. brown-banded cockroach (*Supella longipalpa*) female

    c. German cockroach (*Blattella germanica*) female

    d. oriental cockroach (*Blatta orientalis*) female

2.13 The photo shows an ootheca from a(n)

a. American cockroach (*Periplaneta Americana*)

b. brown-banded cockroach (*Supella longipalpa*)

c. German cockroach (*Blattella germanica*)

d. oriental cockroach (*Blatta orientalis*)

2.14 The photo shows an ootheca from a(n)

a. American cockroach (*Periplaneta Americana*)

b. brown-banded cockroach (*Supella longipalpa*)

c. German cockroach (*Blattella germanica*)

d. oriental cockroach (*Blatta orientalis*)

2.15   The termite soldier shown in the photo is a(n)

    a. eastern subterranean termite (*Reticulitermes flavipes*)

    b. Formosan subterranean termite (*Coptotermes formosanus*)

    c. drywood termite (*Incisitermes spp.*)

    d. none of the above

2.16   The ant shown in the photo is a(n)

    a. big-headed ant (*Pheidole megacephala*)

    b. red imported fire ant (*Solenopsis invicta*)

    c. acrobat ant (*Crematogaster sp.*)

    d. Argentine ant (*Linepithema humile*)

2.17   The ant shown in the photo is a(n)

    a. big-headed ant (*Pheidole sp.*)

    b. red imported fire ant (*Solenopsis invicta*)

    c. acrobat ant (*Crematogaster sp.*)

    d. Argentine ant (*Linepithema humile*)

2.18   The tick shown in the photo is a female

    a. brown dog tick (*Rhipicephalus sanguineus*)

    b. lone star tick (*Amblyomma americanum*)

    c. American dog tick (*Dermacentor variabilis*)

    d. black-legged tick (*Ixodes scapularis*)

2.19   Shown in the photo is a

a. common silverfish (*Lepisma saccharina*)

b. booklouse (Psocoptera)

c. springtail (Collembola)

d. earwig (Dermaptera)

2.20   The wasp shown in the photo is a

a. paper wasp (*Polistes exclamans*)

b. yellowjacket (*Vespula maculifrons*)

c. European hornet (*Vespa crabro*)

d. European paper wasp (*Polistes dominula*)

2.21 **The wood-boring beetle shown in the photo is a**

   a. false powderpost anobiid beetle

   b. true powderpost lyctic beetle

   c. buprestid beetle

   d. cerambycid beetle

2.22 **Match the images of adult droppings with the type of rat.**

   a. House mouse

   b. Roof rat

   c. Sewer/Norway rat

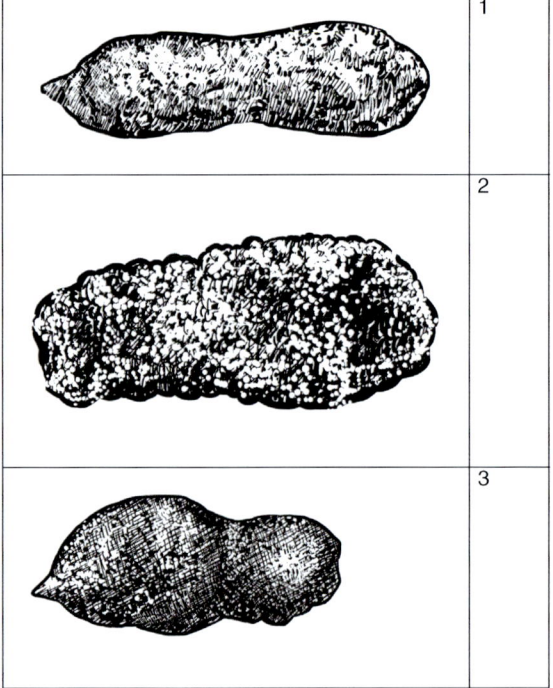

3 Mosquitoes

Mosquitoes are responsible for transmitting deadly diseases. They are in the order Diptera, meaning they have two wings; however, they are different from the filth breeding flies, by not having three segmented antennae and sponging lapping mouthparts.

The most important mosquitoes are the malaria mosquitoes (*Anopheles gambiae* and *An. arabiensis* of Africa), the encephalitis vectors (*Culex pipiens* (northern house mosquito) and *Cu. quinquefasciatus* (southern house mosquito)), and the yellow fever and dengue virus vectors (*Aedes aegypti* (yellow fever mosquito) and *Ae. albopictus* (Asian tiger mosquito)).

Mosquitoes have a holometabolous life cycle with an egg, larva, pupa, and adult stage. All the immature stages are aquatic, and the adult is terrestrial.

Mosquitoes develop in a wide variety of aquatic habitats. *Anopheles* and *Culex* mosquitoes usually occur in locations of permanent or semi-permanent water, such as pools, lakes, streams and water containers. These water sources usually have a large open water surface. The mosquitoes will usually fly about 1–5 miles from their larval development areas to obtain a blood meal. *Aedes aegypti* and *albopictus* mosquitoes occur in smaller, temporary water containers, puddles, leaf axils and tree holes. Other *Aedes*, called floodwater mosquitoes, can develop in salt marshes that contain brackish water. Although mosquitoes develop in a wide variety of habitats, they do not develop in pure sea water.

Mosquitoes are annoying with their buzzing sound, and a single bite can be irritating. When mosquitoes bite, they inject saliva into the skin that can elicit an allergic reaction. For many disease vectors, the saliva may contain pathogens that can infect the host and cause disease.

Mosquito control is achieved best by using integrated mosquito management. It starts with minimizing bites through the use of window screens on houses and animal shelters, and the use of treated bed nets and

© Partho Dhang, Philip Koehler, Roberto Pereira and Daniel D. Dye II 2022. *Key Questions in Urban Pest Management: A Study and Revision Guide* (P. Dhang et al.) DOI: 10.1079/9781800620179.0003

personal repellents. Habitat modification is a reliable method that eliminates adult resting sites and source reduction of larval development areas. Biological control is usually implemented for control of larval populations using predators (e.g. top feeding minnows and copepods) or pathogens and pathogen products (i.e., *Bacillus thuringiensis israeliensis* and *B. sphericus*). Genetic control is being implemented by releasing sterilized male mosquitoes (genetically modified, radiation sterilized, or transgenic mosquitoes). Several methods of chemical control are usually implemented. Larvicides using insect growth regulators or surface films can be applied to larval development areas. Adulticides can be applied to adult resting sites to provide residual control. Resistance to adulticides has been a problem for some mosquito species, like *Anopheles* and *Aedes* disease vectors.

Key Questions

3.1   What is the family name for mosquitoes?

    a. Muscidae

    b. Stratiomyidae

    c. Culicidae

    d. Chironomidae

3.2   What habitat is the location(s) where mosquito larvae live?

    a. Terrestrial

    b. Aquatic

    c. Arboreal

    d. Marine

3.3   Which mosquitoes suck blood?

    a. Females

    b. Males

    c. Toxorhinchites

    d. Both males and females

3.4   When do *Aedes aegypti* and *Aedes albopictus* mosquitoes usually bite?

    a. Nighttime

    b. Daytime

c. Sunrise

d. Midnight

3.5 **Which sex of mosquito has plumose antennae?**

a. Males

b. Females

c. None of these

3.6 **Which pestiferous mosquito larval species does not obtain air through a siphon tube at the water surface?**

a. *Aedes*

b. *Culex*

c. *Anopheles*

d. *Coquellettidia*

e. Both *Anopheles* and *Coquellettidia*

3.7 **Which group of mosquitoes lay their eggs in a raft on the water surface?**

a. *Anopheles*

b. *Culex*

c. *Aedes*

d. *Psorophora*

3.8 **What is the food for mosquito larvae?**

a. Algae, organic matter, and protozoa

b. Fish

c. Clams and shrimp

d. Turtle feces

3.9 **How long after adult emergence do mosquitoes mate?**

a. 1 hour

b. 12 hours

c. 1 day

d. 7 days

3.10 **How long does it take a female to develop and lay eggs after a blood meal?**

    a. 1 day

    b. 3 days

    c. 7 days

    d. 2 weeks

3.11 **What is the term for mosquitoes exuding a blood-like liquid after biting and getting a full blood meal?**

    a. Diarrhoea

    b. Honeydew production

    c. Pre-digestion

    d. Pre-diuresis

3.12 **Which mosquito is considered an urban mosquito and is capable of transmitting the Zika, dengue, yellow fever and Chikungunya viruses?**

    a. *Culex pipiens (quinquefasciatus)*

    b. *Anopheles quadrimaculatus*

    c. *Aedes albopictus*

    d. *Aedes taeniorhynchus*

3.13 **Which mosquito breeds in standing water with high organic content and is one of the main vectors of West nile virus?**

    a. *Culex pipiens (quinquefasciatus)*

    b. *Anopheles quadrimaculatus*

    c. *Aedes albopictus*

    d. *Aedes taeniorhynchus*

3.14 **Which mosquito develops in permanent water and was the main vector of malaria in the US?**

    a. *Culex pipiens (quinquefasciatus)*

    b. *Anopheles quadrimaculatus*

c. *Aedes albopictus*

d. *Aedes taeniorhynchus*

3.15 **Which mosquito is mostly a nuisance and not the prime vector of disease pathogens?**

a. *Culex pipiens (quinquefasciatus)*

b. *Anopheles quadrimaculatus*

c. *Aedes albopictus*

d. *Aedes taeniorhynchus*

3.16 **What was the main reason that malaria was eradicated from Florida and the US?**

a. DDT

b. Draining swamps

c. Exclusion

d. Insecticide treated bed nets

3.17 **Which would be great places for breeding *Aedes aegypti* and *Aedes albopictus* mosquitoes?**

a. Piles of discarded tires

b. Swamps next to houses

c. Salt marshes

d. Flowing rivers

3.18 **Which was one of the original methods of controlling mosquito larvae in water?**

a. Geraniol applications

b. Methoprene briquets

c. Oil applications

d. Bti *(Bacillus thuringiensis israeliensis)*

3.19 **Which is a type of bacteria that can be used to kill mosquito larvae?**

a. Bti *(Bacillus thuringiensis israeliensis)*

b. Methoprene

c. Geraniol

d. Temephos

3.20 **Insect juvenile hormone mimics are used for control of mosquito larvae. What stage of mosquito is usually killed by insect growth regulators?**

a. Egg

b. Larva

c. Pupa

d. Adult

3.21 **Which is an insect growth regulator used for mosquito control?**

a. Methoprene

b. Naled

c. Malathion

d. Bti *(Bacillus thuringiensis israeliensis)*

3.22 **ULV or ultralow volume insecticide applications are used to kill**

a. mosquito larvae

b. mosquito flying adults

c. resting mosquito adults

d. mosquito eggs

3.23 **The optimal size for an ULV droplet should be**

a. <1 micron

b. 1–5 microns

c. ~50 microns

d. 20–30 microns

3.24 **Mist blowers are used to provide mosquito control. What are the places that should be treated with mist blowers?**

a. Sides of structures and vegetation

b. Roads and sidewalks

c. Streams and ponds

d. All of the above choices

3.25 Biological control of mosquito larvae has been accomplished with the release of

a. methoprene

b. naled

c. mosquitofish and copepods

d. birds

3.26 When doing mosquito control work, what is a type of PPE that is not standard for other types of pest control?

a. Face respirator

b. Repellent

c. Gloves

d. Boots

3.27 Which repellent, around since the second world war, is considered the most effective for protection from biting mosquitoes?

a. Geraniol

b. Spearmint

c. Diethyl toluamide

d. Citronella

3.28 Which is a mosquito attractant that is often used to improve mosquito catch in light traps?

a. Octenol

b. Ethanol

c. Methanol

d. Butyric acid

3.29 Ovitraps have been used to catch, monitor, and control mosquitoes. What is the purpose of an ovitrap?

a. Capture mosquito eggs

b. Capture egg-laying mosquitoes

c. Count larvae that hatch from mosquito eggs

d. Produce more mosquitoes

3.30 Which is the most common method for monitoring mosquito larvae?

a. Dumping

b. Sneaking up on larvae

c. Collecting aquatic plants

d. Dipping

4 Bed Bugs

Bed bugs are bloodsucking insects that consume mainly human blood, but may also suck blood from other animals, such as birds and bats. Both sexes, including adults and nymphs, feed on blood. Bed bugs are normally nocturnal and will feed on their hosts when the host is asleep. To maintain the blood flowing and prevent coagulation, the bed bugs inject saliva into the wound, which causes the skin to itch and become swollen.

Bed bugs are hemimetabolous insects that have similar forms as nymphs and adults. The adults do not develop true wings so adults and nymphs are similar in shape, except that the adults' cuticle is dark brown as opposed to the light brown to beige color of the nymphs. Adults are approximately 6 mm long, as nymphs vary from less than 1 mm, for first instars, to approximately adult size for the 5th-instar nymphs. The newly hatched bed bugs are almost colorless and similar to the adult except they are much smaller. When full of blood, the bed bug body becomes swollen, and the color changes to dark red. Fully-engorged adults have a fat cigar-shaped body, but body shape will soon return to a more normal appearance as the blood is digested and the excess water is eliminated. As the bed bug eliminates feces, the insect returns to its normal shape.

Bed bugs are mostly nocturnal insects that hide in cracks and other locations during the day, leaving their refuges at night in order to feed. Bed bugs have a peculiar odor that can be smelt in locations where the infestation is high, and which serves as an attractant to other bed bugs. Populations are usually found in agglomerations. As these bed bugs defecate the digested blood and deposit pheromones, places with heavy infestations acquire a particular bed bug smell.

Bed bug infestations typically start when the insects (adults, nymphs, or mixed populations) are brought into a structure through clothes, bedding, furniture, suitcases, or even on human hosts. Bed bugs are difficult to kill with insecticides since they only take blood as their diet, and their anatomical features minimize the pickup of pesticides from surface deposits. Dust formulations are normally more efficient for elimination of bed bugs than liquid residuals.

Modern populations of bed bugs are known to be resistant to insecticides, especially those in the pyrethroid group. Because populations of bed bugs were treated heavily with DDT and related compounds in the past, these treatments may have provided the basis for the cross resistance to DDT and pyrethroids.

Bed bugs harbor close to where people sleep, including bed frames and mattresses. They are often difficult to control. Restrictions on the use of certain products on locations or surfaces exposed to humans makes control challenging. Prevention is certainly the best way to avoid bed bugs, but an integrated management program using different products and tools can guarantee their elimination.

Key Questions

4.1   In order to distinguish the common bed bug from the tropical bed bug one should compare

    a. the lateral extensions of the pronotum, which almost touch the insect eyes in the tropical bed bug

    b. the lateral extensions of the pronotum, which almost touch the insect eyes in the common bed bug

    c. the length of the antennae, which is two times longer in the tropical bed bug

    d. the length of the antennae, which is two times longer in the common bed bug

4.2   The order, family and genus of bed bugs are

    a. Heteroptera, Camacidae, *Bedbugae*

    b. Heteroptera, *Cimex*, *lectularius*

    c. Hemiptera, Cimicidae, *Cimex*

    d. Homoptera, Cimicoidea, *lectularius*

4.3 **The scientific names for the common bed bug and the tropical bed bug are, respectively**

    a. *Cimex hemipterus* and *Cimex lectularius*

    b. *Cimex columbarius* and *Cimex hemipterus*

    c. *Cimex lectularius* and *Cimex columbarius*

    d. *Cimex lectularius* and *Cimex hemipterus*

4.4 **Bed bugs are**

    a. ectoparasites that suck the blood of humans only

    b. endoparasites that suck the blood of humans and other mammals

    c. endoparasites that live in bird digestive tracts

    d. ectoparasites that suck the blood of humans, other mammals and birds

4.5 **Bed bugs feed on**

    a. animal blood for their entire life

    b. on bed bug secretions as an early instar nymph (1st instar), and then blood for the rest of their lives

    c. on their mother's *pseudomilk* as a nymph and then blood as adults

    d. on bat or bird blood as nymphs and on human blood as adults

4.6 **Bed bugs**

    a. can survive long periods (>1 year) without feeding

    b. can survive only short periods (<1 month) without feeding

    c. cannot survive without feeding at least 2 times a week

    d. cannot survive without feeding at least 4 times a month

4.7 **Adult bed bugs, when fully developed, can reach**

    a. 10 cm in body length

    b. 1 mm in body length

    c. 5–6 mm in body length

    d. 1 cm in body length

4.8   In relation to body size, male bed bugs are

    a. larger than the females

    b. the same size as the females

    c. smaller than the females

    d. longer, but skinnier than the females

4.9   Bed bug eggs can be recognized as

    a. oval white structures <1 mm in diameter attached to the substrate

    b. red round structures deposited in piles on wood furniture

    c. pinkish brown spheres glued to the bed frame

    d. long skinny black structures deposited on the walls of water-containing vessels

4.10   How many eggs will a bed bug female lay in her lifetime in conditions of good nutrition?

    a. 10–30

    b. 30–40

    c. 400–500

    d. 3000–6000

4.11   How many nymphal instars do bed bugs go through?

    a. 3

    b. 5

    c. 6–8

    d. >10

4.12   When plenty of feeding opportunities are available, each bed bug nymphal instar lasts approximately

    a. 1 day

    b. 3 days

    c. 7 days

    d. 115 days

4.13 **Bed bug meals represent what percent of bed bugs' body weight?**

   a. 150–600%

   b. 15–60%

   c. 600–2000%

   d. 10%

4.14 **When feeding on human hosts, bed bugs usually**

   a. crawl into the skin folds

   b. prefer to bite without crawling onto the host

   c. prefer to bite on the portion of the head covered by hair

   d. crawl into body openings

4.15 **Because of the bed bug biting habits, several bed bug bite marks usually appear to be in a**

   a. row

   b. circle

   c. triangle

   d. zig-zag pattern

4.16 **Bed bugs are known to transmit which diseases?**

   a. Malaria and Zika

   b. AIDS and cancer

   c. Flu and rubella

   d. None

4.17 **Compared to cockroach legs, the bed bug's legs**

   a. have no soft pads so bed bugs can climb smooth surfaces more easily

   b. have no soft pads so bed bugs cannot climb smooth surfaces

   c. have many soft pads so bed bugs can climb smooth surfaces better than cockroaches

   d. have many soft pads that make it impossible for bed bugs to climb smooth surfaces

4.18 **The two species of bed bugs that are commonly found feeding on humans are**

    a. the common bed bug, *Cimex hemipterus*, and the tropical bed bug, *Cimex lectularius*

    b. the tropical bed bug, *Cimex hemipterus*, and the common bed bug, *Cimex lectularius*

    c. the common bed bug, *Cimex comunales*, and the tropical bed bug, *Cimex topicalius.*

    d. the temperate bed bug, *Cimex temperatus*, and the tropical bed bug, *Cimex tropicalius*

4.19 **Despite not transmitting any diseases, bed bugs' medical importance is due to**

    a. secondary reactions and infections at the site of bites due to the salivary protein injected into the human body when bed bugs feed

    b. potential for development of anaemia in humans due to severe blood loss in locations with large bed bug populations

    c. presence of allergens in the cast skins and defecations left by bed bugs in infestation sites

    d. all of the above

4.20 **The common bed bug feeds on**

    a. humans and other animals

    b. human hosts only

    c. only on female humans over 50 years of age

    d. human hosts and other mammals

4.21 **Bed bugs are attracted by**

    a. CO_2 only

    b. human body odor only

    c. CO_2 and human body odor and warmth

    d. just body warmth

4.22 **Bed bug females can lay**

    a. only one egg per day

    b. more than three eggs per day on average, depending on nutrition

    c. four eggs per week

    d. hundreds of eggs per week

4.23 **The number of eggs deposited by a female bed bug depends on**

    a. the number of matings the female has had

    b. the nutritional state of the female bed bug

    c. the number of males the female has mated with

    d. the time of year

4.24 **The type of mating between a male and a female bed bug is called**

    a. "Traumatic courtship" because the males drag the females before mating

    b. "Benevolent mating" because the males regurgitate blood for the females to ingest

    c. "Traumatic insemination" because the male punctures the female's abdomen to deliver the sperm inside the female's body

    d. "Benevolent insemination" because the male allows the female to rest during mating

4.25 **When a bed bug is disturbed, it releases**

    a. the "Bed Bug Stink", consisting of (E)-2-octenal and CO_2

    b. the "Bed Bug Gas" consisting of (E)-2-hexanal and chitin

    c. the "*Cimex* alarm pheromone" consisting of (E)-2-octenal and (E)-2-hexenal

    d. the "*Cimex* disturbance signal" consisting of chitin and palmitic oil

4.26 **Bed bugs are**

    a. more sensitive to high heat than drywood termites and flour beetles

    b. not sensitive at all to high temperatures below 130°C

c. less sensitive to high heat than drywood termites and flour beetles

d. killed immediately by temperatures of 25°C and above

4.27 During heat treatments, bed bugs may escape death by

a. passing through wall penetrations, cracks and other openings into unheated areas

b. seeking refuge in areas within a heated room where the heat does not penetrate easily

c. seeking cool spots where heat cannot easily penetrate

d. all of the above

4.28 Populations of the bed bug, *Cimex lectularius*, are very hard to control with certain pesticides because

a. these insects do not drink the pesticides applied to structures

b. species of *Cimex* are naturally immune to synthetic pesticides

c. many *Cimex lectularius* populations are resistant to pyrethroid pesticides

d. only gaseous pesticides can affect bed bugs

4.29 The presence of an active bed bug colony in a location can be detected by

a. presence of living eggs, nymphs or adults

b. the presence of cast skins and egg shells

c. the characteristic bed bug smell

d. the presence of bed bug fecal spots

4.30 Some signs of the presence of bed bugs in a location may include

a. light colored fecal spots from the digestion of insect cuticle

b. dark fecal spots that result from the defecation of digested blood

c. several parts of insects consumed by the bed bugs

d. cylindrical dark brown color feces containing insect parts

5 Flies

Flies belong to the order Diptera and have only one pair of wings. The hind pair of wings are reduced to be club-shaped balancing organs. Apart from houseflies, the most important flies to the urban pest management industry are the filth-breeding flies. The filth-breeding flies are considered very important disease transmitting flies throughout most parts of the world.

Filth-breeding flies are usually broken down by the pest management industry into large flies and small flies. The large flies are strong fliers and can be considered invaders into residences and food handling establishments. Many of the large flies are prevalent around livestock and poultry production facilities. Other important sources of large flies are dead animals and waste facilities like dumpsters and landfills. Some of the large flies of importance are the house fly, blow fly and flesh fly.

Small filth-breeding flies are not considered strong fliers. They usually develop inside or near residences and food-handling establishments. Small flies are sometimes called gnats and develop in places like drains, residues of organic matter and food waste. The ones of importance are the Phorid (humpbacked) fly, Drosophila vinegar (fruit) fly, drain fly and fungus gnat.

It is important to understand the conditions where filth-breeding fly larvae develop in order to control them in urban settings. The food source for filth-breeding fly larvae is usually decaying plant and animal matter. The most common fly is the house fly, and its larvae develop in farm animal manure and decaying plant material. With livestock and poultry production being concentrated into small areas near urban centers, house flies can develop in manure piles and migrate into residences and commercial establishments. The house fly can also develop in decaying garbage and trash containers. Other fly larvae, like blow flies and flesh flies, develop in decaying meat, dead animals, and high protein decaying waste. Even if there is no

© Partho Dhang, Philip Koehler, Roberto Pereira and Daniel D. Dye II 2022. *Key Questions in Urban Pest Management: A Study and Revision Guide* (P. Dhang et al.)
DOI: 10.1079/9781800620179.0005

decaying meat or dead animals, these fly larvae can develop in carnivorous animal waste (dog and cat feces) and invade structures.

The ability of adult flies to move rapidly from decaying plant and animal materials make them important mechanical disease vectors. They pick up disease organisms on their bodies or orally, and then move them to human food by contact with their contaminated bodies or by regurgitation of pathogens. The important pathogens mechanically transmitted by flies are food poisoning bacteria (*Shigella, Salmonella, E. coli*) and cholera. Hundreds of pathogens have been associated with flies. Fly control is important to prevent the mechanical transmission of pathogens.

Key Questions

5.1   What are the two general groups of filth-flies that are found in food handling facilities?

   a. Small and large flies

   b. Long and short flies

   c. Tall and low flies

   d. Long legged and short legged flies

5.2   What are the two groups of flies based on their habitat and feeding behavior?

   a. Sucking and biting flies

   b. Chewing and lapping flies

   c. Sponging and lapping flies

   d. Filth and biting flies

5.3   The most famous and important filth-breeding fly is the

   a. soldier fly

   b. stable fly

   c. house fly

   d. blow fly

5.4 Garbage collection in urban areas is usually _____ to prevent house fly problems.

    a. daily

    b. weekly

    c. monthly

    d. biweekly

5.5 What is the type of mouthpart on a house fly?

    a. Piercing sucking

    b. Lapping

    c. Siphoning

    d. Sponging lapping

5.6 What kind of fly maggots have been seen migrating from the walls of houses?

    a. House fly

    b. Stable fly

    c. Bottle fly or blow fly

    d. Soldier fly

5.7 What fly does not lay eggs on decaying flesh, but deposits first stage larvae?

    a. Flesh fly

    b. Bottle fly

    c. House fly

    d. False house fly

5.8 Which fly is about a third of an inch long and is much larger than a house fly? It has three stripes on the thorax and a checkerboard pattern on the abdomen.

    a. Bottle fly

    b. Soldier fly

    c. Lesser house fly

    d. Flesh fly

5.9 Which fly is sometimes seen migrating from the base of the toilet into the house?

    a. Bottle fly

    b. Soldier fly

    c. Lesser house fly

    d. Flesh fly

5.10 Which is a small fly with a humped back that runs rapidly along surfaces?

    a. Cheese skipper

    b. Flesh fly

    c. House fly

    d. Phorid fly

5.11 This fly breeds in any decaying organic matter of high protein content, and the larvae are often found in tiny cracks of food handling and processing equipment.

    a. Drosophila fly

    b. Moth fly

    c. Phorid fly

    d. Fungus gnat

5.12 These small flies breed in overwatered plants in offices and apartments.

    a. Fungus gnats

    b. Drosophila flies

    c. Drain flies

    d. Moth flies

5.13 What is the common name for Drosophila flies?

    a. Humpbacked flies

    b. Soldier flies

    c. Fungus gnats

    d. Fruit or vinegar flies

5.14 **Why do supermarkets often have fans blowing over their fresh produce and vegetables?**

    a. Their customers often are hot and need to be cooled

    b. Fruit flies do not fly in moving air

    c. The moving air disperses the smell of rotting produce and vegetables

    d. The moving air cools the produce and vegetables

5.15 **What type of antenna do house flies and blow flies have?**

    a. Aristate

    b. Stylate

    c. Plumose

    d. Pilose

5.16 **What type of antenna do tabanid flies (horse flies and deer flies) have?**

    a. Stylate

    b. Plumose

    c. Pilose

    d. Plumose

5.17 **What type of mouthpart does a horse fly have?**

    a. Piercing sucking

    b. Chewing

    c. Piercing lapping

    d. Siphoning

5.18 **What type of mouthpart does a stable fly have?**

    a. Piercing sucking

    b. Chewing

    c. Piercing lapping

    d. Siphoning

5.19 **What is the preferred breeding location for a moth or drain fly?**

    a. Cattle dung

    b. Horse dung

    c. Straw or silage on the ground

    d. Sewers and drains

5.20 **Fruit flies or vinegar flies feed on which type of food?**

    a. Fruit

    b. Vegetables

    c. Yeast

    d. Fungus

5.21 **Insect light traps (ILTs) are often used to control flies indoors in commercial food establishments. How should they be placed?**

    a. Indoors and face the outside entrance doorways

    b. Outside to capture flies before they enter the establishment

    c. Inside near entrances but not facing the entrance doorways

    d. Inside in closets and areas that are dark all the time

5.22 **How far apart should ILTs (insect light traps) be positioned in a food handling establishment?**

    a. 5–10 ft

    b. 10–15 ft

    c. 20–25 ft

    d. 40–50 ft

5.23 **What type of device is most frequently used to kill flies in an ILT (insect light trap)?**

    a. Glue board

    b. Electrocution grid

    c. Fly cage (trap)

    d. Pheromone cage

5.24 **ULV sprays are often used to control house flies. What does ULV stand for?**

a. Under the light value

b. Ultralow volume

c. Uber land view

d. University U-Lever

5.25 **House flies are known to be resistant to many insecticides. Why do they develop so much resistance?**

a. Insecticides have been used extensively to control house flies

b. House flies develop rapidly from egg to adult

c. House flies do not migrate long distances

d. All of these

5.26 **What size of screens are usually used on windows and doors to keep flies from entering houses through doors and windows?**

a. 20 x 20 mesh

b. 14 x 14 mesh

c. 18 x 16 mesh

d. 15 x 15 mesh

5.27 **What is the fly pheromone in the active ingredients statement on some fly bait labels?**

a. Muscadoom

b. Z-9-tricosene

c. Imidacloprid

d. Dinotefuran

5.28 **Can some fly baits be used indoors?**

a. Yes

b. No

5.29 **What kind of fly could be most easily controlled by changing a household plant's watering schedule?**

a. Moth fly (psychodid)

b. Humpbacked fly (phorid)

c. Fruit fly (Drosophila)

d. Fungus gnat (sciarid or fungivorid)

5.30 **Stable flies are often a problem for outdoor events, like weddings and horse-riding competitions. What is a solution for control of these flies?**

a. Empty dumpsters

b. Spread and dry hay and silage associated with animal manure

c. Clean up trash and garbage

d. Remove manure from the area

6 Cockroaches

Cockroaches can be found in virtually all habitats. There are cockroaches that live in temperate and tropical forests, grasslands, salt marshes, aquatic habitats, caves, and deserts. Most people interact with cockroaches in structures, ships, aircraft, and dwellings. Cockroaches are mostly crepuscular and are most active shortly after sunset and before sunrise.

Cockroaches are characterized by a flattened and broadly oval body. The development of the cockroach is hemimetabolous, meaning that there is an egg, nymphal, and adult stage. The eggs of cockroaches are usually deposited in an egg capsule called an ootheca. Ovoviviparous cockroaches form an egg capsule externally from the body but then withdraw it into a brood sac, females provide water until the eggs hatch, and the nymphs emerge from the female's body. For viviparous cockroaches, the ootheca is withdrawn into the female's body, the eggs develop in the brood sac, and eggs are fed nutrients by the female until the birth of the nymphs. Most of these novel developments in reproduction were to protect eggs from parasitoids and predators.

Cockroach nymphs and adults are primarily scavengers living on feces, decaying leaves and wood, as well as dead animals. In structures, they have been found feeding on soap, glue and wire insulation, but they usually feed on human food scraps.

Cockroaches have many adverse effects on people. They can cause psychopathology where the thought or sight of cockroaches, or contact with surfaces where cockroaches have been, can affect a person's perception of their own well-being. They can cause mechanical damage by staining paper and fabric with their feces or oral secretions. Cockroaches are even known to eat book bindings, glue and starched fabric. They are the most important cause of allergies and asthma for inner city children. Cockroaches can

© Partho Dhang, Philip Koehler, Roberto Pereira and Daniel D. Dye II 2022. *Key Questions in Urban Pest Management: A Study and Revision Guide* (P. Dhang *et al.*)
DOI: 10.1079/9781800620179.0006

mechanically transmit pathogens in the move from sewers to food preparation surfaces or consume partially decayed food scraps.

Due to the problems cockroaches cause, control is important. The German cockroach is the most difficult cockroach to control worldwide due to its resistance to many insecticides. As a result, cockroach Integrated Pest Management (IPM) is the best method of protecting people and their properties from cockroaches. IPM starts with prevention. Infestations are often introduced or maintained by movement of infested supplies, storage containers, boxes, and pallets. These items need to be inspected to prevent the establishment of infestations. Cockroaches need food, water, and harborage to survive and thrive. Sanitation and cultural control involve the removal of food, water, and harborage that cockroaches rely on in order to survive.

Key Questions

6.1   What part of the leg of the cockroach is responsible for their ability to climb smooth surfaces and also take up residual insecticides into their body?

   a. Tibial spines

   b. Tarsal pads (plantanulae) and arolium

   c. Tarsal claws

   d. Coxa

6.2   Where are cockroach eggs laid and what is the major way they are killed?

   a. Oothecae and wasp parasitoids

   b. The soil and bird predation

   c. Cracks and spider predation

   d. On walls and smashing by humans

6.3   Cockroaches can mainly detect air movement, sound, and surface vibration with what part of their body?

   a. Antennae

   b. Legs

   c. Abdomen

   d. Cerci

6.4   Fat bodies in cockroaches are used for

a. lipid storage

b. urate storage

c. bacteria harborage

d. all of the above

6.5   The cockroach digestive system is separated into the foregut, the midgut, and the hindgut. Which of these is responsible for allowing food to pass into the body of the cockroach for nutrition?

a. Foregut

b. Midgut

c. Hindgut

d. All of the above

6.6   The Malpighian tubules connect to the gut at which place?

a. Hind midgut, just before the pyloric valve

b. Rectum

c. Hindgut

d. At the esophageal valve

6.7   Cockroaches are in the order Blattodea. What other insect group is also placed in the same order with them?

a. Mantids

b. Grasshoppers

c. Crickets

d. Termites

6.8   Cockroaches are known to aggregate due to chemical depositions. What is the source of these deposits?

a. Oral regurgitates

b. Feces

c. Antennal glands

d. Abdominal glands

6.9   About how many species of bacteria of public health import-
       ance have been collected and identified from cockroaches?

       a. 5

       b. 10

       c. 100

       d. 500

6.10   Cockroaches are associated with allergies and asthma.
       What group of allergens have been documented to cause
       allergic responses in people?

       a. Peri a

       b. Bla g

       c. Peri au

       d. Blatt o

6.11   What type of vacuum should be used to remove cockroaches
       and their debris from an infested location?

       a. HEPA

       b. Standard canister or upright

       c. Shop vacuum

       d. Backpack

6.12   Cockroaches can cause medical issues related to physical
       damage to humans. What is the most common problem
       medical professionals encounter?

       a. Cockroach ingestion by people

       b. Cockroach bites

       c. Cockroaches lodged in ear canals

       d. Squashed cockroaches causing skin irritation

6.13   What is the first step in German cockroach courtship?

       a. Mutual antennal fencing (antennation)

       b. Male wing raising

c. Mounting

d. Female calling

6.14 **Although cockroaches aggregate, they also have to disperse. What initiates cockroach dispersal from a harborage?**

a. Light

b. Sound

c. Salivary compounds

d. Starvation

6.15 **Which cockroach is spread mainly by human cockroach dispersal?**

a. Oriental cockroach

b. American cockroach

c. Surinam cockroach

d. German cockroach

6.16 **Cockroaches cause psychological stress and are associated with filth and disease. What is true about cockroaches?**

a. Cockroaches are not associated with filth

b. Cockroaches are disliked across many cultures and geographic regions

c. Cockroaches do not cause anxiety, social isolation, and sleeplessness in people with psychological distress

d. Most people do not try to eliminate cockroach infestations

6.17 **What is the lowest temperature that allows cockroaches to survive?**

a. −100°C

b. 0°C

c. 20°C

d. −20°C

6.18 **What part of the body do hissing cockroaches use to produce sound?**

a. Antennae

b. Mouth

c. Anus

d. Spiracles

6.19 **Where did the American cockroach originate?**

a. America

b. Africa

c. Asia

d. Australia

6.20 **Clothing damage by cockroaches means that cockroaches can digest what nutrients?**

a. Starch and body oils

b. Cotton

c. Wool

d. Polyester

6.21 **What type of feeding is attributed to cockroaches?**

a. Herbivore

b. Carnivore

c. Omnivore

d. Vegetarian

6.22 **When looking at the posterior tip of the cockroach abdomen, what readily differentiates a male and female cockroach?**

a. Presence of styli

b. Presence of cerci

c. Presence of an ootheca

d. Both a and c are correct

6.23 **What is the most prevalent stage of cockroaches encountered in field locations?**

    a. Egg

    b. Nymph

    c. Adult male

    d. Adult female

6.24 **Cockroaches typically forage just after sunset with a lot of locomotor activity. Which cockroach life stages do not follow this pattern?**

    a. Males

    b. Females with egg capsules and 1st–3nd stage nymphs

    c. Late-stage nymphs

    d. Only females with egg capsules

6.25 **Which is true about the social behavior of cockroaches?**

    a. Female cockroaches will eat the male after mating

    b. Males will eat the young to maintain dominance

    c. Females exhibit proctodeal trophallaxis to young nymphs

    d. Nymphs feed food to adult cockroaches

6.26 **What is an albino (white) cockroach?**

    a. Newly molted nymph or adult

    b. A species of cockroach that lives in snow

    c. A cockroach that has been deprived of protein

    d. An adult cockroach that has gone through an additional molt

6.27 **What is true about cockroach locomotion?**

    a. They crawl very slowly from their harborage to food and water

    b. They can only move through large cracks to invade a home

    c. When disturbed, they can accelerate on their hind legs and move very fast

    d. They only move when disturbed

6.28 **Antennae are a very important structure on cockroaches. What is a principal use for them?**

    a. Food determination

    b. Tactile

    c. Smell

    d. Temperature determination

6.29 **How many eyes do cockroaches have?**

    a. Two

    b. Three

    c. Four

    d. Five

6.30 **The nerve cord in the cockroach has three large ganglia. What is the main purpose of those large ganglia?**

    a. Vision

    b. Odor

    c. Locomotion

    d. Reproduction

7 Subterranean Termites

Termites are cellulose (wood) feeding insects. They feed on all types of wood such as dead trees, processed wood, products made of wood, paper, textile, plant roots, litter and soil humus. Termites are social insects, living in a colony with a distinct caste system. In the tropics, termites are considered pests in both agriculture and urban areas. In urban areas they rank top in the list of destructive insects, damaging structures in parts of the United States, Australia, Asia-Pacific and parts of Africa.

Termites are generally categorized into: subterranean termite: termites mostly living in soil and using mud tubes to move above ground; non-subterranean termite or dry wood termite: termites living in moistureless wood with no connection with soil; non-subterranean termite or damp wood termite: termites living in moist wood such as fences, poles or dead trees. They may or may not maintain connection to soil.

Subterranean termites are major pests of structures. The pest enters a structure mainly through a crack in the foundation or floor slab, through wall gaps or expansion joints. They also enter using utility pipes such as electrical and telephone conduits, water and drain pipes. Some species also find above ground routes to enter a structure by forming mud tubes over walls, trees and plants in contact with the structure.

Moisture is a critical factor in subterranean termite survival. At times moisture can become the single reason a termite is attracted to a structure.

Among several species of subterranean termite, *Coptotermes* in the family Rhinotermitidae is known to be an invasive species with the highest economic consequence.

Termite foraging behaviour is not random, but follows a methodical exploration and utilization of the available resources. Decisions to feed or

© Partho Dhang, Philip Koehler, Roberto Pereira and Daniel D. Dye II 2022. *Key Questions in Urban Pest Management: A Study and Revision Guide* (P. Dhang *et al.*)
DOI: 10.1079/9781800620179.0007

not to feed, redistribution of members of workers and soldiers, and formation of subcolonies are all taken based on overall fitness of the colony.

The origin of the termite is from the soil, so treating soil is the most common practice in controlling it. Residual chemicals with repellent and non-repellent properties are the most commonly used chemicals for treatment. Treating the soil, and building foundations and sub-slab areas as an industry practiced intervention method is common to prevent as well as control infestation. The perimeter of a structure is generally treated by soil trenching or rodding/soil injection application techniques.

Another method of treating termite infestation is by the use of termite bait. This has emerged as an alternative to toxic chemical use as well as a rational way to eradicate and suppress colonies. This method is most suitable for treating above ground infestations. Baits using a number of insecticidal active ingredients such as bistrifluron, chlorfluazuron, diflubenzuron, hexaflumuron, noviflumuron, and noviluron are available in the industry.

Key Questions

7.1   Pick the correct match for the following

a. Termite alates have _____.

b. Ant alates have _____.

1. straight antennae, straight waist and equal length wings

2. elbowed antennae, narrow waist and unequal length wings

3. straight antennae, narrow waist, and unequal length wings

7.2   Is it correct that, in termites, the three body parts, namely head, thorax and abdomen, are all joined together as one without constrictions?

a. Yes

b. No

7.3   Away from damaging structures, termites are known for

a. transforming dead wood and other materials containing cellulose into humus

b. producing methane as a result of digestion

c. carrying a number of diseases

d. distribution throughout the world

7.4   **Worker and soldier caste can be easily distinguished by**

a. examining the body size

b. examining the size and structure of the head

c. examining the structure of the antennae

d. the presence and absence of simple eyes

7.5   **Termites are now classified in Blattodea but they used to be under order _____ meaning their fore and hind wings are nearly identical in size and venation**

a. Odonata

b. Orthoptera

c. Isoptera

d. Diptera

7.6   **A mature *Coptotermes formosanus* queen can typically lay the following number of eggs in a day**

a. 250

b. 1000

c. 5000

d. 10,000

7.7   **Is it true that the subterranean termite can construct nests away from soil?**

a. Yes

b. No

7.8   **Termites are considered social insects because they**

a. live in a colony with a large number of individuals

b. have kings and queens

c. share food between themselves

d. have a caste structure performing different tasks

e. all of the above

7.9 **Termite soldiers, where the mandibles are modified as a snout, belong to the genus**

    a. *Heterotermes*

    b. *Nasutitermes*

    c. *Macrotermes*

    d. *Coptotermes*

7.10 **Match the following**

    a. Eastern subterranean termite

    b. Formosan subterranean termite

    c. Asian Mound building termite

    d. Desert subterranean termite

    1. *Heterotermes aureus*

    2. *Macrotermes gilvus*

    3. *Reticulitermes flavipes*

    4. *Coptotermes formosanus*

7.11 **Detection of a carton-like material in an above ground infestation is a sure sign of**

    a. *Nasutitermes sp.*

    b. *Coptotermes formosanus*

    c. *Reticulitermes flavipes*

    d. *Heterotermes aureus*

7.12 **In temperate regions of the world, termites respond to a gradual decrease in temperature by**

    a. moving to the upper layer of the soil

    b. moving closer to each other to generate body heat

    c. moving to the lower layer of the soil

    d. abandoning the nest and moving to a warmer location

7.13 **Termites are divided as a "higher" or "lower" group, which is based on**

a. feeding habit

b. nest construction

c. latitudinal distribution

d. presence and absence of multiple species of bacteria in their gut

7.14 **The subterranean termites considered invasive belong to the following families**

a. Rhinotermitidae, Termitidae and Kalotermitidae

b. Termitidae, Mastotermitidae and Kalotermitidae

c. Rhinotermitidae and Termitidae

d. Kalotermitidae and Hodotermitidae

7.15 ***Coptotermes formosanus* is a native of**

a. Hawaii

b. New Zealand

c. Taiwan

d. Southern China

7.16 **Measuring wood moisture with a moisture meter**

a. is useful for determining the potential for termite attack

b. gives an indication of the success of moisture control programs

c. helps identify the degree of decay

d. indicates quality of ventilation

7.17 **Which one among the following is a critical limiting factor for termite survival?**

a. Moisture

b. Darkness

c. Fungi

d. Soil

7.18 **Primary means of feeding among termites is via _____, which is the mutual exchange of food, fluid and nutrients between colony members.**

a. trophollaxis

b. filter feeding

c. suction feeding

d. fluid feeding

7.19 **The purpose of termites making mud tubes is**

a. concealment while moving

b. providing a moist shelter

c. protecting the termites from ants

d. transporting food

e. all of the above

7.20 **A modest size *Coptotermes formosanus* colony of 350,000 workers can/may consume how much wood per day?**

a. Approximately 1000 g

b. Approximately 500 g

c. Approximately 6 g

d. Approximately 30 g

7.21 **Match the building terms with the description.**

a. Monolithic slab

b. Weep holes

c. Floating slab

d. Termite shield

1. A basic slab type construction where the foundation wall and footing are separated from the concrete slab floor by an expansion joint

2. Openings in mortar between bricks in lower courses to provide drainage for moisture that accumulates between bricks and sheathing

3. A shield (metal) placed in or on a foundation wall, other mass masonry, or around pipes to prevent passage of termites

4. A basic slab type construction where the concrete foundation footing and the slab floor are formed as one continuous unit

7.22   Repellent chemical application done as a perimeter treatment helps to achieve the following:

a. kill all the termites in the vicinity

b. provide a barrier of entry

c. entrap the termites from going back to the colony and kill them

d. all of the above

7.23   Which of the following are characteristics of a repellent termiticide?

a. Contains bifenthrin as an active ingredient

b. Needs to be applied all around the perimeter

c. Allows controlled penetration of termites through the treated zone

d. Recommended to be used as a spot treatment on above-ground infested areas

7.24   Categorize the following active ingredients into repellent or non-repellent.

a. Fipronil   1. Repellent. 2. Non-repellent

b. Cypermethrin   1. Repellent. 2. Non-repellent

c. Imidacloprid   1. Repellent. 2. Non-repellent

d. Chlorfenapyr   1. Repellent. 2. Non-repellent

7.25   Second generation non-repellent termiticides have a unique property of being carried from a poisoned member to other members away from the treatment zone by the process of

a. trophallaxis

b. grooming

c. ingestion

d. all the above

7.26 The term "horizontal transfer" in termite treatment means

a. flow of termiticide from worker to worker

b. flow of termiticide from worker to soldier

c. flow of termiticide from worker to worker and soldier

d. flow of termiticide from worker to reproductives

7.27 Which of the following statements are true for application of in-ground termite bait stations?

a. Needs to be installed between structure and termite colony

b. Needs to be installed between structure and chemical barrier

c. Needs to be installed approximately 3 m apart from each other

d. Needs to be baited only after the termites are intercepted in the station

7.28 A reliable sign of subterranean termite damage in a property is the presence of

a. discarded swarmer wings

b. high moisture levels in the wood

c. hollow sound when wood tapped

d. mud tube on the structure

e. all of the above

7.29 Prevention of subterranean termite infestation in a structure is best achieved by the following methods

a. avoidance of all forms of wood to soil contact

b. destruction of wooden stumps near the structure

c. monitoring of moisture levels in wooden parts of the structure

d. choosing concrete as the building material

7.30 The following materials are resistant against termites

a. fiber reinforced cement

b. gypsum

c. plastic

d. masonry

8 Drywood Termites

Drywood termites live in low moisture wood which serve as both food and shelter for them. These termites do not require ground contact and do not build mud tubes. The colonies are usually small, with numbers ranging from between a few hundred to a few thousand. Unlike subterranean termites, a drywood colony does not need to forage for food; however, with the passing of time the colony can face a serious depletion of food.

Drywood termites were first recognized as a distinct group in the late 19th century and assigned to the family Kalotermitidae. The characteristics of this group are based on the alates (winged primary reproductives) and includes the presence of ocelli, a left mandible with two marginal teeth, 2-segmented cerci, antennae with 10–24 segments, and lack of a fontanelle. However, the common name often associated with this family, "dry wood", is misleading. A more realistic description would be based on its bio-ecology and that would be "single piece nesters".

The colony of the dry wood termite consists of a labyrinth of gallery systems inside wood that expands with time as the population increases. The architecture of the gallery system consists of narrow passageways that inter-connect to a number of larger feeding chambers. Colony fusion between two colonies is known to occur after they contact each other in the same wooden material.

Infestation from dry wood termites usually begins via aerial routes. The mating pair of male and female fly and colonize a wooden structure. This makes any portion of the house constructed with wood vulnerable to attack.

The control of drywood termite starts with visual inspection which is possible when the affected wood is accessible. For inaccessible areas various devices have been developed, adopted from other fields of science. Once infestation is located, spot treatment is undertaken which includes replacing the

© Partho Dhang, Philip Koehler, Roberto Pereira and Daniel D. Dye II 2022. *Key Questions in Urban Pest Management: A Study and Revision Guide* (P. Dhang *et al.*) DOI: 10.1079/9781800620179.0008

infested portion of the wood, subsurface injection with a registered pesticide straight into the galleries, increasing the temperature of the infested wood to at least 50°C and holding it for a period of minimum 1 hour, and fumigation using sulfuryl fluoride (SF).

Fumigation by use of registered fumigants is an efficient and effective method for large infestations. Prevention of infestation from drywood termites is another method of controlling damage. Chemicals used for such work are called wood preservatives.

Key Questions

8.1 **Drywood termites are able to exploit wood with moisture content in the range of**

a. 15–20%

b. 3–12%

c. 1–3%

8.2 **An average mature colony for drywood termites will have individuals in the range of**

a. 10,000–15,000

b. 5000–8000

c. 500–3000

8.3 **Drywood termite infestation is most prevalent in the following climatic zones.**

a. Desert and arid zones

b. Warm humid coastal zones

c. Dry high-altitude zones

d. Dry tropical zones

8.4 **Family Kalotermitidae represents termites that are all "dry wood" in nature.**

a. True

b. False

8.5   How many genera are represented in the family Kalotermitidae?

    a. 21

    b. 25

    c. 35

    d. 40

8.6   Kalotermitids are categorized as invasive in nature due to the

    a. ability to extract maximum moisture from wood

    b. ability to produce viable propagule

    c. ability to resist pesticide

    d. ability to nest in the wood they feed upon

8.7   Drywood termites display "colony fusion".

    a. True

    b. False

8.8   Presence of six-sided compact fecal pellets under a wooden structure is an indication of

    a. subterranean termite infestation

    b. powderpost drywood termite infestation

    c. powderpost beetle infestation

    d. carpenter ant infestation

8.9   Discarded alate wings can be used as a method to distinguish between subterranean and drywood termites by looking at

    a. presence and absence of hairs

    b. pigmentation

    c. number and pattern of veins

    d. length and width ratio

8.10   Is it correct that drywood termites construct mud tubes to travel from one point to another?

a. Yes

b. No

c. Occasionally

8.11   It is correct that the same wood can be infested by multiple drywood termite colonies.

a. True

b. False

8.12   The most commonly used method of detecting a drywood termite infestation is

a. canine detection

b. infra-red camera

c. fiber optic devices

d. visual inspection

8.13   Match the following possibilities.

a. Subterranean termite soldiers have _____

b. Drywood termites have _____

1. pronotum as wide as or wider than the head

2. pronotum of a smaller width than the head

3. mandibles with marginal teeth

4. mandibles devoid of teeth

8.14   The characteristic shape of the drywood termite's fecal pellets is due to the insect's unique ability to

a. feed on low moisture food

b. use them to seal damaged wood

c. conserve metabolic water

d. use it to defend against ants

8.15 **The most common/practical remedial treatments for drywood infestation are**

    a. treatment of galleries with essential oil

    b. fumigation

    c. wood injection with pesticides

    d. wood replacement

8.16 **Drywood termites can be ideally controlled if the infested structure is heated to _____ and the temperature held for a minimum of one hour.**

    a. over 50°C

    b. 40°C

    c. between 35–45°C

    d. over 70°C

8.17 **The commonly used fumigant now available for treating drywood termite treatment is**

    a. phosphine

    b. methyl bromide

    c. sulfuryl fluoride

    d. carbon dioxide

8.18 **Methyl bromide is an effective fumigant against drywood termite but it is restricted due to**

    a. ozone depleting properties

    b. expense

    c. corrosiveness

8.19 **The following are correct for threshold limit value (TLV).**

    a. The values vary depending on the fumigant

    b. It is the air-borne concentration of fumigant to which workers may be repeatedly exposed (8 hours/day, 5 days/week) with no adverse effects

    c. It is not recognized as a reliable measure of worker safety

8.20   Common routes of exposure to fumigants occur through

    a. the nose

    b. the mouth

    c. the skin

    d. the eyes

    e. all of the above

9 Powderpost Beetles and Wood Pests

There are several important beetles that attack wood in nature, but those that cause concern in the urban environment are the ones that are capable of completing their development in seasoned wood, and that can potentially reproduce and produce a new generation of pests that will continuously reinfest wood, causing damage. These pests may be especially difficult to recognize due to their cryptic nature, staying hidden from observation for most of their lifecycle.

Mostly, external signs serve as indications of their presence, but very often, these signs are not obvious until the degree of the damage is considerable. Often exit holes may be the only indication of the presence of a wood infesting pest, and in some cases, the shape and location of these holes can help in the identification of the infesting pest. Another clue for the identification of the pest may be the presence of frass that is left inside the galleries once the insect has exited or died.

Beetles are most often the culprits in damaging wood and mostly individuals belonging to the families Anobiidae, Lyctidae, and Bostrichidae, although others may occasionally be important in different parts of the world. Different species may have preference for different types of wood, and even different ages of wood. Depending on the characteristics of the wood, especially its hardness and the presence of secondary compounds that may enhance or delay the pest development and the damage that it causes to the wood, damage can vary greatly among the different pests. Therefore, a good understanding of the source of the wood or wooden object may be important in the identification of the pest, and potentially the methods to be used in the control of the pest and the reduction of any damage it may cause.

Understanding of all the factors associated with the biology and ecology of the pest is essential for the development of appropriate control and

© Partho Dhang, Philip Koehler, Roberto Pereira and Daniel D. Dye II 2022. *Key Questions in Urban Pest Management: A Study and Revision Guide* (P. Dhang et al.)
DOI: 10.1079/9781800620179.0009

management strategies that will minimize damage and prevent reinfestation of the wood, and potentially prevent the spread of the infestation to other buildings, objects and materials.

Control of these wood pests is not always possible before considerable damage is done. Thus, prevention is important in minimizing damage. Damage that occurs in living wood, or prior to the final use of the lumber, will remain, but is not likely to spread infestations if the wood is properly treated or prepared before use in final products.

Key Questions

9.1   **Wood-boring beetle damage can be distinguished from that of other wood-destroying organisms by**

    a. the presence of small round or oval holes

    b. the meandering, powder-filled larval tunnels in the wood

    c. the six-sided, barrel-shaped frass

    d. a and b above

9.2   **The different families of beetles are known to attack**

    a. only softwoods

    b. softwoods or hardwoods dependent on the beetle family

    c. only hardwoods

    d. both softwoods and hardwoods

9.3   **Certain beetles attack dying trees only and their presence in homes**

    a. requires immediate attention to avoid further damage to the indoor wood

    b. means treatment is unnecessary to prevent further damage

    c. means emerging adults will reinfest the same wood they emerged from

    d. none of the above

9.4   **Because in most wood damaged by beetles, larvae and adults are seldom seen, it is important to**

    a. be able to identify beetle families by examining the wood itself

    b. apply control before identifying the beetle

c. know where the wood was produced

d. know what means of transportation was used to transport the damaged wood

9.5   A 7-year-old wood is considered _____ when dealing with beetle damage identification.

a. old

b. new

c. middle-aged

d. it does not matter for these purposes

9.6   The cut-off point for a wood to be considered "old", as opposed to "new", for the purposes of beetle damage identification is

a. 5 years since the lumber was manufactured

b. 7 years since the lumber was manufactured

c. 10 years since the lumber was manufactured

d. 15 years since the lumber was manufactured

9.7   The designation of a wood as hardwood or softwood

a. has nothing to do with the hardness of the wood

b. is related to the type of tree they come from

c. a and b above

d. none of the above

9.8   Hardwood comes from

a. broad-leaved trees such as cedar and cypress

b. coniferous species such as pine and cedar

c. deciduous species such as oak and ash

d. coniferous trees such as maple and walnut

9.9   Softwood comes from

a. broad-leaved trees such as cedar and cypress

b. coniferous species such as pine and cedar

c. deciduous species such as oak and ash

d. deciduous broad-leaved trees such as maple and walnut

9.10 Hardwoods have

    a. horizontal and vertical resin ducts and no pores

    b. large pores and horizontal and vertical resin ducts

    c. large pores, no resin ducts, and often grain patterns with light-colored sapwood and dark heartwood

    d. no pores and no resin ducts

9.11 Softwoods have

    a. horizontal and vertical resin ducts and no pores

    b. large pores and horizontal and vertical resin ducts

    c. large pores, no resin ducts, and often grain patterns with light-colored sapwood and dark heartwood

    d. no pores and no resin ducts

9.12 The term "powderpost beetle" refers to the fact that

    a. adult beetles only lay eggs on timber previous attacked by other insects which destroy the wood

    b. larvae of these beetles feed on wood reduced to powder by other beetle groups

    c. adult beetles reduce timbers internally to a mass of very fine, powder-like material

    d. Larvae of these beetles reduce timbers internally to a mass of very fine, powder-like material.

9.13 Groups of powderpost beetles include

    a. Cerambicids, Bostrichids, and Lyctids

    b. Anobiids, Bostrichids, and Cerambicids

    c. Lyctids, Anobiids, and Bostrichids

    d. Anobiids, Curculionids, and Carabids

9.14 Anobiid powderpost beetles

    a. are elongated, have a keystone-shaped pronotum, and develop and feed in hardwood only

    b. are elongate-oval with the head deflected downward and develop in softwood or hardwood structures

c. are elongated with parallel-sided wing covers, head deflected downward, and require bark for egg-laying

d. are oval with red covered wings with black dots and develop in bamboo shoots

9.15 Bostrichid powderpost beetles

a. are elongated, have a keystone-shaped pronotum, and develop and feed in hardwood only

b. are elongate-oval with the head deflected downward and develop in softwood or hardwood structures

c. are elongated with parallel-sided wing covers, head deflected downward, and require bark for egg-laying

d. are oval with red covered wings with black dots and develop in bamboo shoots

9.16 Lyctid powderpost beetles

a. are elongate, have a keystone-shaped pronotum, and develop and feed in hardwood only

b. are elongate-oval with the head deflected downward and develop in softwood or hardwood structures

c. are elongated with parallel-sided wing covers, head deflected downward, and require bark for egg-laying

d. are oval with red covered wings with black dots and develop in bamboo shoots

9.17 Lyctid attacks are mostly found in

a. new homes, because the beetles attack hardwoods with a maximum of 3% starch

b. old homes, because the beetles attack softwoods with a minimum of about 3% starch

c. new homes, because the beetles attack softwoods with large pores in which the female can lay her eggs

d. new homes, because the beetles attack hardwoods with a minimum of about 3% starch and large pores in which the female can lay her eggs

9.18 Bostrichids can attack

    a. hardwood only

    b. hardwood and softwood, but not bamboo

    c. hardwood, softwood and bamboo

    d. softwood only

9.19 Cerambycids are also known as

    a. longhorned beetles due to the presence of a horn-like structure similar to that of a rhinoceros

    b. longhorned beetles due to the presence of thin antennae that may be longer than their bodies

    c. waxbeetles due to the presence of waxy structures on the head of the adults

    d. waxbeetles due to the waxy residue left by larvae on the wood after feeding

9.20 Longhorned beetle damage is usually limited to

    a. pine sapwood

    b. oak sapwood

    c. pine heartwood

    d. oak heartwood

9.21 The old house borers are

    a. Bostrichids

    b. Lyctids

    c. Anobiids

    d. Cerambicids

9.22 Conditions that are conducive to wood decay and insect attack include

    a. high moisture and temperatures between 10 and 35°C

    b. low moisture and temperatures between 5 and 15°C

    c. high moisture and cooler temperatures between 5 and 10°C

    d. low moisture and temperatures between 10 and 35°C

9.23 **A common source of wood pests inside homes include**

a. insects flying in through open windows

b. insects brought into the home on infested lumber or furniture

c. insects crawling in under closed doors.

d. insects changing their diets from stored food to wood

9.24 **Most beetles do not develop in wood with a moisture content below**

a. 10–15%

b. 25–35%

c. 40–50%

d. 90–80%

9.25 **Some of the chemicals used in the treatment of wood in order to prevent the development of wood-infesting pests include**

a. sodium chloride and potassium chloride

b. monosaccharides and disaccharides

c. sodium borate and zinc borate

d. sodium bicarbonate and zinc bicarbonate

9.26 **Borate products are recommended for treatment of wood**

a. that is used in any situation in construction

b. in contact with soil, but not exposed to rain

c. exposed to rain, but not to accumulated water

d. out of contact with the ground, and in places protected from liquid water

9.27 **Sealing products such as shellac, varnish, wax and paint prevent beetle attack because**

a. beetles do not like the smell of these products

b. beetles are unable to oviposit if the pores in the wood are filled

c. these products lower the moisture level in the wood

d. these products give the wood a bitter taste

9.28 **Installing a concrete floor in the basement of houses, or the use of plastic barriers between the soil and the wooden structure, can prevent wood-destroying beetle damage because**

a. it provides moisture control, reducing the moisture level in wood below those levels tolerated by wood-destroying beetles

b. it prevents wood-destroying beetles from emerging from the soil

c. it increases the temperature of the wood beyond levels tolerated by wood-destroying beetles

d. it tricks the wood-destroying beetles into thinking they are inside the house already

9.29 **Heat can be used as a non-chemical option for wood-destroying insect treatment as long as the treatment process**

a. gets the air in the structure to a minimum of 50°C (~120°F)

b. lasts for 40 minutes independent of the temperature reached

c. gets at least of 80% of monitoring thermometers (minimum of 5) installed in the structure to 50°C (~120°F)

d. heats all wood in the structure to a minimum of 50°C (~120°F) and that temperature is held for at least 33 minutes

9.30 **Ambrosia beetle damage to wood can be recognized because**

a. their tunnel walls are stained red, purple or black due to their introduction of the ambrosia fungi

b. their tunnels bifurcate several times, and always at 90° angles

c. the wood looks normal except for a slight orange color staining that diffuses for a minimum of 5 cm from the insect gallery

d. the wood is completely destroyed by the presence of the fungus, and crumbles when touched

10 Ants

Ants represent one of the largest groups of insects, and some of the most prevalent pests in households and other human-occupied buildings. Although most ants can bite with their jaws, the ones that cause greater concern are the ones that sting, using a modified ovipositor to inflict pain.

Because ants live in large nests that can house many thousands to millions of individuals, their collective effect is certainly what causes greatest concern as a force that may destroy or consume large quantities of food or other materials important to humans. In agricultural settings, ants contaminate and destroy products and stored foods, but their effect is certainly not diminished in urban environments. In nature, ants perform beneficial functions, preying on pests, aerating soils, moving soil nutrients, and decomposing organic matter, but in urban environments, they can be considered as one of the most destructive urban pests. There are over 12,000 species of ants, and only a dozen or so species are important pests in urban settings.

Ants belong to the order Hymenoptera which also includes bees and wasps, and, like many other hymenopterans, they are social insects with colony duties divided among different castes. Queens conduct the reproductive functions of a colony, while workers take care of the brood and the queens. Ants undergo complete metamorphosis, having egg, larval, pupal, and adult stages. The larvae are immobile and wormlike, and cause no concern as pests. Sterile female workers gather food, feed and care for the larvae, build tunnels, and defend the colony. Males do not participate in colony activities, and their only purpose is to mate.

Ant management requires diligent effort and the combined use of mechanical, cultural, sanitation, and chemical methods of control. Emphasis should be on excluding ants from buildings and eliminating food and water sources.

© Partho Dhang, Philip Koehler, Roberto Pereira and Daniel D. Dye II 2022. *Key Questions in Urban Pest Management: A Study and Revision Guide* (P. Dhang et al.)
DOI: 10.1079/9781800620179.0010

Key Questions

10.1 **The terms monodomous and polydomous refer to**

    a. ant species that live on a single food source or multiple food sources

    b. ant colonies that occupy a single nest site and ant colonies that occupy more than one nest site

    c. ant individuals that express characteristics of a single caste, or characteristics of more than one caste

    d. ant species that have a single caste in the nest or multiple castes in the same nest

10.2 **The terms monogyne and polygyne refer to**

    a. ant species that live on a single food source or multiple food sources

    b. ant individuals that express characteristics of a single gender, or both male and female characteristics

    c. ant species that have a single queen or multiple queens in the nest

    d. ant individuals that have a single function in the nest or multiple functions

10.3 **When ant colonies are formed by "swarming",**

    a. female and male reproductives leave the original nest site, crawl to a new nest location, mate, and start rearing the new workers from eggs laid by the female reproductive

    b. female and male reproductives go on a nuptial flight, mate, and then crawl to a new nest location and start rearing the new workers from eggs laid by the female reproductive

    c. female and male reproductives go on a nuptial flight, mate, and the female starts a new colony by laying eggs and rearing the first brood until workers appear

    d. reproductive females move to a new location with workers that will carry larvae and pupae to start the new nest

10.4   When ant colonies are formed by "budding",

a. female and male reproductives leave the original nest site, crawl to a new nest location, mate, and start rearing the new workers from eggs laid by the female reproductive

b. female and male reproductives go on a nuptial flight, mate, and then crawl to a new nest location and start rearing the new workers from eggs laid by the female reproductive

c. Female and male reproductives go on a nuptial flight, mate, and the female start a new colony by laying eggs and rearing the first brood alone until workers appear

d. reproductive females move to a new location with workers that will carry larvae and pupae to start the new nest

10.5   Ant control should emphasize

a. exclusion of ants from buildings, and elimination of food and water sources

b. use of liquid residual pesticide applications

c. killing foragers

d. killing the ant larvae

10.6   One indirect way to control ants involves controlling

a. butterflies in the garden

b. aphid and scale populations on trees and shrubs that provide sugar solutions to the ants

c. fertilizer nutrients available to the plant so plants do not grow too vigorously

d. shaded areas in the landscape

10.7   In order to control ants that trail on branches of trees, the following method can be applied:

a. a sticky resin or petrolatum applied in a band around the lower part of the trunk of a plant to provide a barrier to the ants

b. use of liquid residual pesticide applications

c. killing foragers

d. killing the ant larvae

10.8   Ant control should emphasize use of

    a. mechanical methods to kill ants

    b. bait products that will be taken to the nest and spread among the ant population

    c. residual liquid pesticides applied to hard surfaces

    d. residual powder pesticides applied to the soil surface

10.9   Ants can be kept out of buildings by

    a. placing mango tree leaves on top of the refrigerator

    b. caulking cracks and crevices around foundations

    c. opening windows to increase air flow

    d. sprinkling salty water on door landings

10.10   To prevent ants from entering buildings, it is better to

    a. plant bamboo and fruit trees close to the building

    b. plant ant-attracting plants close to the building so ants stay on these

    c. avoid having plants that attract ants near buildings

    d. keep windows closed all the time

10.11   One can control ants indirectly by

    a. applying a mixture of sugar and ethanol to the corners of rooms

    b. applying salty water on door frames

    c. praying for afternoon rains that normally kill ants

    d. applying systemic insecticides to control sucking insects

10.12   In relation to ant control, potted plants

    a. cannot serve as food or shelter for ants

    b. can serve as food or shelter for ants

    c. are never a place pest ants will be found

    d. have no role in ant survival or biology

10.13 Chemical control of ants is most successful when the insecticide treatment is focused on

    a. foraging ants on kitchen counters

    b. queens and larvae inside nests

    c. alate females that will form new nests

    d. the few male ants that will fertilize the queen or queens

10.14 The most efficient way to get a toxicant into the nest to kill the ant colony is by

    a. using baits that are taken into the nest

    b. using insecticides to kill foragers on their way out to food sources

    c. using insecticides to kill foragers on their way back to the nest

    d. using dry powder formulations that stick to the feet of the foragers

10.15 Most adult ants cannot

    a. ingest liquid food, therefore liquid baits will not work in controlling ants

    b. process solid food, therefore granular bait particles never work against adult ants

    c. suck liquids from a solid matrix, therefore only solid baits work against ants

    d. ingest solid particles, therefore solid food is given to the larvae which can process solid food

10.16 In an ant nest, individuals with wings are

    a. intruders that came into the nest to rob the food storage

    b. special workers that fly long distances in order to forage for fresh food

    c. reproductive males and females that will go on a nuptial flight before the females start new nests

    d. a different species of ants that nests in the same location in order to get protection

10.17 **The process for exchanging food between different members of the ant society is called**

a. trophozooid

b. trophollaxis

c. tropicallis

d. tropholysis

10.18 **The ideal ant bait incorporates three elements including**

a. fast-acting, repellent toxicant on a fermented food basis

b. fast-acting, non-repellent toxicant on a fermented food basis

c. slow-acting, non-repellent toxicant on a preferred food basis

d. slow-acting, repellent toxicant, on a preferred food basis

10.19 **Besides being broadcast, liquid and solid baits can also be used in**

a. spray applicators

b. swirl bait applicators

c. power-spray equipment

d. designated bait stations

10.20 **Broadcast applications of ant baits outdoors should not be made if**

a. there is no prediction of rain within the next 2 hours

b. there is a prediction of rain within the next 2 hours

c. the bait will be carried by the pest

d. the pest is within 5 m of an occupied building

10.21 **Baits are most effective when ant colonies are searching for food**

a. early in spring, before flowering plants begin to bloom, because ants will begin feeding on nectar as soon as it is available

b. late in the fall, after flowering plants drop their bloom

c. in the hottest part of summer, because ants like to forage in extreme heat

d. in the winter, because ants are a cold-loving insect

10.22 In order to avoid ant infestations, sanitation actions should include

  a. rinsing clothes with anti-ant soap

  b. washing windows so the glass reflects the solar light and deters the ants from approaching

  c. rinsing empty soft drink containers and removing them from the building

  d. washing floors with anti-ant soap

10.23 A material that can be used in order to protect certain areas from foraging ants is

  a. a black material because ants dislike this color

  b. a bitter bait because ants can taste the bitterness through air

  c. a sweet material because ants do not like sweet materials

  d. a sticky material because ants will get stuck and stop foraging

10.24 A cheap and safe alternative for ant control is

  a. sugar water, which is repellent to ants

  b. soapy water, which is effective in controlling foraging ants in a building

  c. cold water, which kills ants on contact

  d. soda water, which ants will drink and make them explode

10.25 Ant baits are most effective

  a. in the summer when the weather is really hot

  b. in the winter when ants forage the most

  c. early in the spring when ants are foraging vigorously for food

  d. in the fall, when the cooling weather forces them to eat more

10.26 Active ingredients used in ant baits should be

  a. slow-acting because the pesticide will make the ants suffer more

  b. slow-acting because the pesticide can be widely distributed in the nest before it starts killing the ants that have consumed the poison

  c. fast-acting so the applicator can see the ants dying right away

  d. fast-acting so the ants can all die while the pest-management personnel can see the ants dying

10.27 **For ants that can be nesting both indoors and outdoors, the ant control program should start with**

    a. bait application indoors, followed by outdoor baiting

    b. blocking ant escape routes from the house

    c. use of broadcast application of insecticides indoors

    d. blocking ant access into the building

10.28 **Thief ant infestations should be treated with**

    a. sugar-based bait because thief ants only eat sweet materials

    b. oil-based bait because they prefer greasy foods

    c. protein-based bait because they do not eat sugar or oils

    d. something other than sugar-based bait as it is not very effective with thief ants because they prefer greasy foods, meats, cheeses and similar animal products

10.29 **When controlling pharaoh ants inside large buildings, sometimes control methods such as applying sprays may appear to be working but in reality**

    a. they are feeding the colonies and making them stronger

    b. they are killing other ants that look like pharoah ants

    c. they are fragmenting pharoah ant colonies into smaller, more dispersed colonies

    d. they are washed off by excessive moisture indoors

10.30 **Quarantine programs aimed at preventing the spread of fire ants (*Solenopsis invicta* and *Solenopsis richteri*)**

    a. have worked well as can be seen by the restricted number of locations where these ants exist

    b. have not been implemented anywhere, therefore the ants have spread worldwide

    c. have caused the spread of other pest ants

    d. have been ineffective and this ant species has become invasive in many parts of the world

11 Fleas, Ticks and Mites

Fleas, ticks, and mites are important ectoparasites on humans. They feed on human and animal blood and tissues. Fleas are insects with six legs, three body regions, and antennae on the head; whereas ticks and mites are arachnids with eight legs in the adult stage, two body regions, and lack antennae. All three groups are important transmitters of disease and are difficult to control.

Fleas

Fleas are insects in the order Siphonaptera with holometabolous development. The order is characterized by the body being laterally flattened so the flea can adeptly move through the fur and hair of the host. The hind legs are adapted for jumping. The most important fleas are associated with cats, dogs, rodents and poultry.

Flea control is most commonly accomplished with oral or topical applications of drugs or insecticides to pets as prescribed by veterinarians. These applications control fleas by either contact with the host or ingestion of blood containing a toxicant. Pest control technicians often control fleas with residual sprays containing a mixture of insecticides and insect growth regulators.

Ticks

Ticks are in the arthropod order Acarina. The two important families of ticks are the soft ticks (Argasidae) and the hard ticks (Ixodidae).

Ticks are prevented with protective clothing, personal repellents, and acaricides applied to infested habitats. Pest control technicians usually treat the insides of houses and yards around houses focusing on the edges of woods, lawns, and borders adjacent to known tick habitats.

© Partho Dhang, Philip Koehler, Roberto Pereira and Daniel D. Dye II 2022. *Key Questions in Urban Pest Management: A Study and Revision Guide* (P. Dhang *et al.*) DOI: 10.1079/9781800620179.0011

Mites

Mites are small arthropods with a cephalothorax and four pairs of legs as nymphs and adults. First stage larval mites have six legs. The complete life cycle can occur in as little as 2–3 weeks, and huge populations can occur rapidly. Most mites in urban areas are parasites associated with rodents, birds, or humans: chiggers, stored food mites and house dust mites.

Key Questions

11.1 **What is the species of flea that is the main pest of cats and dogs in the US?**

    a. *Ctenocephalides felis felis* – cat flea

    b. *Ctenocephalides canis* – dog flea

    c. *Pulex irritans* – human flea

    d. *Xenopsylla cheopis* – oriental rat flea

11.2 **What is the species of flea that is the main vector of plague throughout the world?**

    a. *Ctenocephalides felis felis* – cat flea

    b. *Ctenocephalides canis* – dog flea

    c. *Pulex irritans* – human flea

    d. *Xenopsylla cheopis* – oriental rat flea

11.3 **What is the species of flea that is associated with wild animals, like raccoons and possums?**

    a. *Ctenocephalides felis felis* – cat flea

    b. *Ctenocephalides canis* – dog flea

    c. *Pulex irritans* – human flea

    d. *Xenopsylla cheopis* – oriental rat flea

11.4 **Which is a term for the infestation of the skin by fleas burrowing and surrounded by irritated skin tissue?**

    a. Cat scratch fever

    b. Flea allergy dermatitis

    c. Tungiasis

    d. Plague

11.5   **What is a term for the loss of hair and skin irritation caused by a flea infestation?**

a. Cat scratch fever

b. Flea allergy dermatitis

c. Tungiasis

d. Plague

11.6   **A blocked flea gut is a result of infection with what disease?**

a. Cat scratch fever

b. Flea allergy dermatitis

c. Tungiasis

d. Plague

11.7   **Cat and dog tapeworms are transmitted between pets by what means?**

a. Ingestion of dog and cat feces

b. Flea larvae ingesting tapeworm eggs and the fleas being ingested by the host

c. Scratching the tapeworm eggs into the skin

d. Fleas injecting tapeworm eggs into skin with their saliva

11.8   **How are cat flea eggs laid for larval development?**

a. Flea eggs are glued to the hairs of a host and larvae fall off into the habitat

b. Fleas leave the host and lay eggs in the environment

c. Fleas retain eggs until they hatch, then larvae fall off into the habitat

d. Fleas do not glue eggs to hairs or fur; they fall off the host and then hatch in the habitat

11.9   **What is the main flea larval food?**

a. Decaying vegetation

b. Adult flea feces (dried blood)

c. Hair and fur from the host

d. Skin of the host

11.10 **Fleas pupate by spinning a silken cocoon. How long can fleas remain in the cocoon as pre-emerged adults waiting for a host?**

    a. 1 week

    b. 1 month

    c. 6–12 months

    d. 2 years

11.11 **How would you treat a room infested with pre-emerged adult fleas (inside the cocoon)?**

    a. Spray the entire floor area

    b. Wait until the fleas stop emerging as adults

    c. Spray only the locations where the pet slept

    d. Don't bother spraying until adults are out of the cocoon; it protects the fleas. Stimulate the fleas to get them out of the cocoon, and then spray

11.12 **Which is not transmitted by fleas?**

    a. Plague (*Yersinia pestis*)

    b. Murine typhus (*Richettsia typhae*)

    c. Cat scratch disease (*Bartonella henselae*)

    d. Dog/cat heartworm (*Dirofilaria immitis*)

11.13 **What is the organ located on the first pair of legs of a tick that is used for detecting hosts by odor?**

    a. Hallers organ

    b. Stigmata

    c. Hypostome

    d. Chelicerae

11.14 **Which stage of ticks has six legs?**

    a. Egg

    b. Larva

    c. Nymph

    d. Adult

11.15 **What are the body regions of ticks and mites called?**

    a. Head and thorax

    b. Head and abdomen

    c. Head and mouthparts

    d. Head and cephalothorax

11.16 **Ticks are outstanding vectors of pathogenic organisms. Select the key factors that allow them to successfully transmit pathogens.**

    a. Wide host range and tendency to feed on several hosts during their lifetime ensures ample opportunity to acquire and transmit pathogens

    b. High reproductive potential ensures large populations and a high frequency of host contact

    c. They feed slowly and attach to the host for relatively long periods. This allows sufficient time for pathogen acquisition and transmission as well as vector dispersal by means of host movement

    d. All of these

11.17 **What is a main characteristic of hard ticks that differentiates them from soft ticks?**

    a. Long mouthparts

    b. A scutum that covers the entire back of male ticks

    c. A scutum that covers the entire back of both male and female ticks

    d. Short legs

11.18 **Which tick is able to cause tick paralysis?**

    a. American dog tick (*Dermacentor variabilis*)

    b. Brown dog tick (*Rhipicephalus sanguineus*)

    c. Black-legged or deer tick (*Ixodes scapularis*)

    d. Lone star tick (*Amblyomma americanum*)

11.19 Which tick is the main vector of Lyme disease bacteria, *Borrelia burgdorferi*?

a. American dog tick (*Dermacentor variabilis*)

b. Brown dog tick (*Rhipicephalus sanguineus*)

c. Black-legged or deer tick (*Ixodes scapularis*)

d. Lone star tick (*Amblyomma americanum*)

11.20 Which tick stage is usually responsible for transmitting Lyme disease pathogens to humans?

a. Egg

b. Larva

c. Nymph

d. Adult

11.21 The life cycle of black-legged ticks includes how many hosts?

a. One

b. Two

c. Three

d. Four

11.22 Which tick is capable of completing its entire life cycle in a house or structure?

a. American dog tick (*Dermacentor variabilis*)

b. Brown dog tick (*Rhipicephalus sanguineus*)

c. Black-legged or deer tick (*Ixodes scapularis*)

d. Lone star tick (*Amblyomma americanum*)

11.23 Brown dog ticks can lay many eggs in a house or structure, usually under furniture or in cracks. Since they are hidden, what is the problem?

a. Thousands of seed ticks

b. Staining from the eggs

    c. Staining from the blood consumed by the female tick

    d. Dead female ticks that have died after laying eggs

11.24 Which mite causes respiratory distress in humans?

    a. Dust mite (*Dermatophgoides* spp.)

    b. Straw itch mite (*Pyemotes tritici*)

    c. Human itch mite (*Sarcoptes scabei*)

    d. Northern fowl mite (*Ornythonyssus sylvarum*)

11.25 Which mite is involved with skin irritation between the webbing of fingers and skin folds of the wrists?

    a. Dust mite (*Dermatophgoides* spp.)

    b. Straw itch mite (*Pyemotes tritici*)

    c. Human itch mite (*Sarcoptes scabei*)

    d. Northern fowl mite (*Ornythonyssus sylvarum*)

11.26 Which mite is a parasite/predator of insects?

    a. Dust mite (*Dermatophgoides* spp.)

    b. Straw itch mite (*Pyemotes tritici*)

    c. Human itch mite (*Sarcoptes scabei*)

    d. Northern fowl mite (*Ornythonyssus sylvarum*)

11.27 Which mite is a parasite and sucks the blood of humans in infested houses?

    a. Dust mite (*Dermatophgoides* spp.)

    b. Straw itch mite (*Pyemotes tritici*)

    c. Human itch mite (*Sarcoptes scabei*)

    d. Northern fowl mite (*Ornythonyssus sylvarum*)

11.28 Which is the larger arthropod?

    a. Tick

    b. Mite

11.29 **Which arthropod has three body regions?**

    a. Tick

    b. Mite

    c. Flea

    d. Spider

11.30 **The mouth opening of ticks and mites occurs on which body part?**

    a. Chelicerae

    b. Hypostome

    c. Scutum

    d. Stomata

12 Sporadic Pests

Sporadic pests are usually insects or other arthropods that people infrequently encounter or that live and breed outside and then enter a building in large numbers. Many times, people are not familiar with these pests and think they may cause significant damage to their structures or endanger them by stinging or biting. At times, customers will encounter these pests, making life interesting and challenging for pest management professionals. Although these sporadic pests may be "minor" pests to the pest management technician, they can be considered a huge problem to the customer.

The key to solving problems with sporadic pests is to determine the source and cause of the problem. Many sporadic pests originate in vegetation. Lady beetles feed on aphids and mealy bugs and are beneficial in controlling plant pests. However, they can become very numerous and enter structures in large numbers during the fall of the year. Other insects like thrips, leafhoppers, brown marmorated stink bugs, boxelder bugs and many other beetles that invade structures are huge problems for customers, but very difficult to prevent or control.

Other sporadic pests originate in decaying organic matter on the ground. Many of these are rather slow moving, but invade in huge numbers. For instance, millipedes have mass migrations from their breeding grounds when weather or soil conditions change. Some aquatic insects have massive emergences. Mayflies and aquatic midges are known to completely cover buildings close to water.

Then there is the odd pest such as plaster beetles. Some of the Hymenoptera are sporadic pests too. Cicada killer wasps, umbrella wasps, and velvet ants are examples of ones that people may encounter in urban areas.

© Partho Dhang, Philip Koehler, Roberto Pereira and Daniel D. Dye II 2022. *Key Questions in Urban Pest Management: A Study and Revision Guide* (P. Dhang *et al.*) DOI: 10.1079/9781800620179.0012

Key Questions

12.1 **What is the most common lady beetle to enter houses in the US?**

    a. Two spotted lady beetle (*Adalia bipunctata*)

    b. Multicolored Asian lady beetle (*Harmonia axyridis*)

    c. Convergent ladybug (*Hippodamia convergens*)

    d. Sevenspotted lady beetle (*Coccinella septempunctata*)

12.2 **What do plaster beetles feed on?**

    a. Plaster

    b. Paint

    c. Fungus

    d. Sugar

12.3 **When sticky traps are placed outside on the ground around structures, what is the most frequently collected insect?**

    a. Springtails

    b. Cockroaches

    c. Earwigs

    d. Centipedes

12.4 **Which insect has wings that have hairs rather than membranous wings?**

    a. Mosquito

    b. House fly

    c. Mayfly

    d. Thrips

12.5 **Which aquatic insect has four membranous wings that are folded vertically over its body when resting on surfaces?**

    a. Mayfly

    b. Caddisfly

    c. Stonefly

    d. Aquatic midge

12.6   Which aquatic insect looks like a mosquito, but doesn't bite or feed at all in the adult stage?

   a. Mayfly

   b. Caddisfly

   c. Stonefly

   d. Aquatic midge

12.7   Why do moth caterpillars migrate from outside locations and into houses?

   a. Looking for an overwintering location

   b. In search of tree and bush leaves to eat

   c. In search of a place to build a cocoon and pupate

   d. Avoiding birds

12.8   Some caterpillars sting people. How do these caterpillars inflict their painful sting?

   a. A stinger on the tip of their abdomen

   b. A bite from their mandibles

   c. Urticating hairs on their bodies

   d. Large pinchers on their front legs

12.9   What are millipedes usually called by customers?

   a. Thousand leggers

   b. Hundred leggers

   c. Pillbugs

   d. Sowbugs

12.10   What do centipedes eat?

   a. Decaying leaves and vegetation

   b. Leaves on plants

   c. Stems of plants

   d. Small animals in their environment

12.11 **What is the class and order for sowbugs and pillbugs?**

    a. Arthropoda, Chilopoda

    b. Crustacea, Isopoda

    c. Arthropoda, Insecta

    d. Crustacea, Amphipoda

12.12 **What are the class and order for lawn shrimp?**

    a. Arthropoda, Chilopoda

    b. Crustacea, Isopoda

    c. Arthropoda, Insecta

    d. Crustacea, Amphipoda

12.13 **What is a common Hemipteran to invade houses from surrounding vegetation?**

    a. Leaf-footed bug

    b. Water strider

    c. Brown marmorated stink bug

    d. Bed bugs

12.14 **What is a brightly red and black colored true bug (Hemipteran) that can invade structures in large numbers?**

    a. Boxelder bug

    b. Brown marmorated stink bug

    c. Leaf footed bug

    d. Blood-sucking conenose bug

12.15 **All species of scorpions that invade houses in the US are considered highly dangerous.**

    a. True

    b. False

12.16 **What is the description of a scorpion?**

    a. Four pairs of legs, one pair of large pedipalps, seven trunk segments and six segments for the tail

    b. Seven pairs of legs, no large pedipalps, no trunk and 13 segments for the tail

  c. One pair of antennae, three pairs of legs and 14 segments for the tail

  d. No antennae, three pairs of legs, and 14 segments for the tail

12.17 Which arthropod defends itself with acetic acid?

  a. Scorpion

  b. Tarantula

  c. Whip scorpion

  d. Centipede

12.18 What is a large spider that many people think is highly venomous?

  a. Scorpion

  b. Tarantula

  c. Whip scorpion

  d. Centipede

12.19 A customer brings firewood into the house and stores it next to the fireplace. Tiny cylindrical beetles are seen afterwards in the house. What is a likely identification of those beetles?

  a. Longhorned wood borers

  b. Powderpost beetles

  c. Ground beetles

  d. Bark beetles

12.20 Which wingless wasp runs around on the ground, looks like a large ant and is covered with pubescence hairs?

  a. Carpenter ant

  b. Umbrella wasp

  c. Velvet ant

  d. Fire ant

12.21 **Where would you look for the larvae of Dermestid beetles found in a house?**

a. Kitchen cabinet

b. Bedroom

c. Stuffed animals and hides

d. Bathroom

12.22 **What fabrics would clothes moths typically consume?**

a. Nylon

b. Rayon

c. Polyester

d. Wool

12.23 **Do Buprestid beetles reinfest wood in a structure?**

a. Yes

b. No

12.24 **Which insect appears on a 17 or 13 year cycle and makes a lot of noise?**

a. Migratory locust

b. Cricket

c. Periodical cicada

d. Tree frog

12.25 **Which bug can transmit Chagas disease and is a blood sucking insect?**

a. Conenose bug (Triatomine)

b. Bed bug (Cimicid)

c. Wheel bug (Arilus)

d. Boxelder bug (Rhopalid)

12.26 **Why would leafhoppers enter a structure?**

a. Looking for an overwintering location

b. In search of tree and bush leaves to eat

c. In search of a place to build a cocoon and pupate

d. They are attracted to lights at night

12.27 Where do cicada killer wasps choose to build their nests?

a. Lawns and bare soil

b. Eaves of buildings

c. In trees

d. In wood

12.28 Where do carpenter bees build their nests?

a. Lawns and bare soil

b. Walls of buildings

c. In trees

d. In wood

12.29 What is the description of a paper wasp nest?

a. A hole in the ground

b. A hole in wood

c. A comb of cells with no covering

d. A comb of cells surrounded by a paper-like covering

12.30 Why do lady beetles enter houses?

a. To avoid water

b. To obtain water

c. To find shade

d. To overwinter

13 Stored Product Pests

Plant and animal products that can remain in storage for long duration are categorized as stored products. These include plant products that are used as food such as corn, rice, millets, legumes, nuts, dried fruits and spices; and non-food such as fabric and tobacco. Animal products under this category can also be food and non-food products such as dried meat, fish, dried dog food and leather.

Stored products also include processed or finished food items such as flour, biscuits, noodles, chips and crackers; and non-food items such as cigars and products made from leather and fur.

A specific type of insect attacks and damages stored products, and these are known as stored product pests. These pests reduce the nutritional quality of food products by eating the richer parts of seeds and grains and introducing pathogens. They cause contamination by leaving body parts and feces in the food products. They also destroy and damage fabric and leather. Overall, these insects are considered major economically important pests.

Stored product insects are comprised of only two insect groups or orders. These insect groups are moths (Lepidoptera) and beetles (Coleoptera) which also comprises weevils.

The most common target for stored product pests is grains and grain-derived products. This is a cause for concern due to their importance in global food security, so considerable effort is put into controlling these pests.

Stored product pests on grains are broadly classified based on their nature of feeding. A stored product pest may enter a commodity at any point, from its origin to final use. This can start from picking and harvesting, then to its packing and transportation, storage in warehousing, its use in processing mills and food manufacturing factories, and eventually placement in retail stores and household pantries.

© Partho Dhang, Philip Koehler, Roberto Pereira and Daniel D. Dye II 2022. *Key Questions in Urban Pest Management: A Study and Revision Guide* (P. Dhang *et al.*) DOI: 10.1079/9781800620179.0013

Apart from food-based stored products, non-food such as fabric in the form of textiles, wool, carpet, etc. are also damaged by pests. The major pests are cloth moths including the case making cloth moth and the webbing cloth moth; and beetles belonging to the Dermestidae family such as the carpet beetle, the ladder beetle and the hide beetle. These insects feed on keratin and protein remnants in the product in the presence of certain fungi.

Control and management of stored product pests can be achieved by a number of methods. These include a step-by-step process that covers sanitation, proper storage, safe transportation, insecticide application and continuous monitoring.

Key Questions

13.1 Which of the following items could be infested by stored product pests?

a. Wooden furniture

b. Books and paper

c. Biscuits and chips

d. Carpets

13.2 Common invertebrate stored product pests are all insects.

a. False

b. True

13.3 The majority of stored product pests belong to which of the following insect orders?

a. Coleoptera and Lepidoptera

b. Isoptera and Hemiptera

c. Diptera and Psocoptera

d. Orthoptera and Hymenoptera

13.4 Mite infestation is characterized by

a. formation of slow-moving mite dust on the infested surface

b. a characteristic minty odor when crushed

c. discoloration of the infested food

d. dispersal aided by birds, rodents and humans

13.5 **Match the common name of the stored product pest with its correct scientific name.**

a.	Rice weevil	1	Tribolium confusum
b.	Red flour beetle	2	Sitophilus oryzae
c.	Cigarette or tobacco beetle	3	Stegobium paniceum
d.	Drug store beetle	4	Anthrenus flavipes
e.	Carpet beetle	5	Lasioderma serricorne

13.6 _____ **is identified by the presence of six saw-like projections on each side of the thorax.**

a. *Tribolium confusum*

b. *Oryzaephilus surinamensis*

c. *Stegobium paniceum*

d. *Sitophilus oryzae*

13.7 **Which beetle is found to damage products made out of animal skin, bird feathers and all types of animal hair?**

a. Tobacco beetle

b. Drug store beetle

c. Red flour beetle

d. Carpet beetle

13.8 **Which of the following beetles makes a hole in the grain, deposits its eggs and covers them with a gelatinous substance?**

a. *Oryzaephilus surinamensis*

b. *Tribolium castaneum*

c. *Sitophilus oryzae*

d. *Lasioderma serricorne*

13.9 **The sight of dense webbing, silk strands, and prominent size in milled cereals is an indication of _____ infestation.**

a. Tobacco beetle

b. Carpet beetle

c. Indian meal moth

d. Drug store beetle

13.10 The stored product pest that is capable of boring through seed coats to penetrate inside is a

    a. saw toothed grain beetle

    b. confused flour beetle

    c. lesser grain borer

    d. rice weevil

13.11 Unlike rice and granary weevils, pea and bean weevils are not true weevils

    a. true

    b. false

13.12 Which of the following moths show limited flight?

    a. Cloth moth

    b. Indian meal moth

    c. Angoumois grain moth

13.13 The confused flour beetle and the red flour beetle can be most easily distinguished by

    a. examining the antennae

    b. measuring the body length

    c. checking the body colour

    d. all of the above

13.14 Match the following pest to its correct description.

    a. Confused flour beetle

    b. Indian meal moth

    c. Rice weevil

    d. Drug store beetle

    f. Cloth moth

    1. Buff, tan or golden colored moth with a wing span of 1.5 inch that is capable of flight

    2. Colorful moth with characteristic grey and coppery wings with dark bands and capable of flight

3. Reddish-brown beetle with snout, four light red or yellow spots on the wing covers

4. Reddish-brown beetle about 3 mm long and antennae end in a four-segmented club

5. Reddish brown, oval and body covered with tiny hairs to give it a golden sheen

13.15 Fabric pests look for the following for their survival:

a. starch

b. fat

c. glue

d. protein

13.16 Fill in the blanks. Bakeries predominantly use milled flour and other milled ingredients; these face damage from _____. Breweries or snack food manufacturers that store whole grain or kernel ingredients are mostly at risk from _____.

a. external feeders

b. internal feeders

13.17 Globally the largest volume of stored product pest management work is undertaken on the following items:

a. beans and pulses

b. grains and cereals

c. nuts and spices

d. tobacco and animal food

13.18 The best way to keep a check on stored product pest in large establishment is by the use of _____ and _____ .

a. inspection

b. insect light traps

c. insecticide application

d. sanitation

13.19 **A distinct foul odor on infested food products caused by the flour beetle is due to**

a. insect feces

b. fungi

c. quinone compounds

d. decomposition

13.20 **Yellow color glue boards are hung around stored product warehouses as they are proven to _____.**

a. control moisture

b. attract and capture pests

c. repel pests

13.21 **The most common fumigant used for grain products is _____ .**

a. methyl bromide

b. sulfuryl fluoride

c. phosphine

d. carbon dioxide

13.22 **Which among the two fumigants has the following characteristics?**

a. High grain penetration – 1. Phosphene 2. Methyl Bromide 3. Both

b. Toxicity to target insects – 1. Phosphene 2. Methyl Bromide 3. Both

c. Easy to transport – 1. Phosphene 2. Methyl Bromide 3. Both

d. Easy to apply – 1. Phosphene 2. Methyl Bromide 3. Both

13.23 **In order to increase the efficiency of the treatment, _____ can be used in combination with phosphine for effective pest control.**

a. pesticides

b. carbon dioxide

c. heat

d. light traps

13.24 Which of the following statements is correct for fumigants?

a. Fumigants are gases that act as a respiratory poison

b. Fumigant products are labeled under Toxicity Category I

c. Fumigants have better efficacy at lower temperatures

d. Fumigants are effective against all types of grain product pests

13.25 Psocid infestation are often not controlled by phosphine fumigation.

a. True

b. False

13.26 The best way to prevent stored products pest occurrence in a warehouse is by

a. proper aeration of the product

b. storing the product in dark conditions

c. preventing moisture reaching the stock

d. removing cracks and crevices from the flooring

13.27 When an infestation is found in a supermarket the first action should be to

a. remove the infested material

b. spray a registered pesticide

c. install light traps

d. file a complaint against the supplier

13.28 Which of the following are the correct use of insect light traps

a. Should not be located directly over exposed products

b. Should face out towards the exit

c. Should be installed at a height not exceeding 8 feet

d. Require periodic checks for UV-A output.

13.29 Which of the following insects is not attracted to light and is incapable of flight?

a. Granary weevil

b. Drug store beetle

c. Red flour beetle

d. Indian meal moth

13.30 The ideal temperature used commercially to kill insects in the food processing industry is

a. 35°C

b. 50°C

c. 60°C

d. 80°C

14 Rodents

Rats and mice are among the most common vertebrate pests found in urban environments, specifically in buildings. They are troublesome because they are well-adapted to live alongside people but destroy or contaminate food and fabrics, cause structural damage, damage wires, which increases the risk of fires, and are also carriers of disease organisms.

Managing rodent pests requires greater skills and knowledge because rodents are more intelligent than other urban pests, being able to recognize and avoid control devices and products. An integrated and persistent approach is necessary in order to circumvent rodent intelligence and learning capacity. Understanding their habits and food preferences is a key component in management of the pest population.

Rats eat and contaminate food products such as fruit, vegetables, grains and meat, along with many packaged products such as pet food. Rodents are also responsible for the destruction, damage or staining of textiles, upholstery, printed material and insulation, especially when these materials are used as bedding in nests. Rats create holes in walls, doors and windows, and can gnaw on electrical wiring, water pipes and gas lines, causing great damage and even fires.

Another problem caused by rodents are from their fecal droppings, urine and hairs which are left all over the area the rodents occupy. These excretions or droppings have been associated with several human health issues.

Rats enter buildings in several ways but require openings of 2.5 cm or larger. They can also chew holes through wooden window or door frames. Rats have poor vision, but they compensate with well-developed senses of smell, taste, hearing and touch. They use these senses to locate food and avoid danger, foraging for food mostly during the night to avoid encounters with people and other dangers. Rats are agile and quick, and can climb and

© Partho Dhang, Philip Koehler, Roberto Pereira and Daniel D. Dye II 2022. *Key Questions in Urban Pest Management: A Study and Revision Guide* (P. Dhang *et al.*)
DOI: 10.1079/9781800620179.0014

swim well, and squeeze through small openings. They are extremely wary of new items or situations and may take several days to adjust to changes in their environment before investigating new items such as food or nesting materials.

Key Questions

14.1 The Norway rat is also known as

    a. black rat or roof rat

    b. brown rat or sewer rat

    c. white rat or wharf rat

    d. lab rat or north rat

14.2 The roof rat is also known as

    a. yellow rat, vegetable rat or apple rat

    b. house rat or brown rat

    c. black rat, fruit rat or citrus rat

    d. south rat or day rat

14.3 The scientific name of the Norway rat is

    a. *Rattus northus*

    b. *Rattus norvegicus*

    c. *Rattus rattus*

    d. *Rattus lefthus*

14.4 The scientific name of the roof rat is

    a. *Roofus rattus*

    b. *Communis rattus*

    c. *Toppus rattus*

    d. *Rattus rattus*

14.5 From head to the tip of the tail, the average Norway rat measures approximately

    a. 20 cm (8 inches)

    b. 40 cm (16 inches)

    c. 120 cm (50 inches)

    d. 12 cm (5 inches)

14.6 **Because the Norway rat and the black rat are about the same size, a good way to differentiate the two species is by comparing the tails because**

    a. the black rat's tail is shorter than the body, while the Norway rat's tail is longer than the body

    b. the black rat's tail curls up away from the body, while in the Norway rat the tail curls down

    c. the black rat's tail is 2× the body length, while in the Norway rat the tail is 3× the body length

    d. the black rat's tail is longer than the body, while the Norway rat's tail is shorter than the body

14.7 **Do Norway rats and black rats coexist in the same area?**

    a. No, these rats never share the same area of habitat, despite having similar size and level of aggressiveness

    b. Yes, but only in cold climates

    c. Yes, despite the fact that Norway rats are much more aggressive and stronger than black rats

    d. No, because their preferred temperatures are very different

14.8 **The litter sizes for both the Norway rat and roof rat are normally**

    a. from 6 to 12 young

    b. from 2 to 4 young

    c. from 20 to 30 young

    d. from 12 to 15 young

14.9 **In the wild, the normal lifespan for Norway rats is**

    a. about 2.5 years

    b. shorter than 1 year

    c. less than 3 months

    d. just 3 weeks

14.10 Rats become sexually mature at

a. 3 weeks after birth

b. 10 weeks after birth

c. 3 months after birth

d. 8 months of age

14.11 Rats are most active

a. during the hour after sunset or before sunrise

b. during the warmest portion of the day

c. from 11 PM to 1 AM

d. from 4 to 6 PM

14.12 Norway rats normally nest in

a. ground burrows or natural and man-made cavities

b. open-air nests protected from rain

c. only in burrows the rats dig themselves

d. nests of other animals, using areas not occupied by these animals

14.13 Besides nesting in trees, roofs and attics, the roof rats may also nest in

a. wall voids

b. underground burrows and under debris

c. voids in equipment

d. all of the above

14.14 Rats are considered "opportunistic omnivores" which means they eat

a. a restricted diet of few food items only

b. different types of food they find in the environment

c. a vegetarian diet

d. a diversity of animal products only

14.15 Rats are most active searching for food

    a. in the late morning

    b. in the late evening

    c. at dusk and just prior to dawn

    d. from noon to 3 in the afternoon

14.16 Water consumption in rats

    a. is not important since they do not drink liquid water and get all the water they need from food

    b. is restricted to less than 5 ml of water per day for an adult rat

    c. greater than half a liter per day for an adult rat

    d. depends on their activity level, climatic factors, the food they consume and age

14.17 Urban rat populations

    a. normally do not migrate unless resources become scarce

    b. migrate 2–4 times a year to different locations

    c. migrate every 3–5 weeks to avoid predators

    d. migrate every winter to a new location

14.18 Indoors, the house mouse breeding season lasts approximately

    a. 3 months

    b. 6 months

    c. 9 months

    d. all year

14.19 Under ideal situations such as indoors, with abundant food and lack of predators and other dangers, the lifespan of a mouse can be

    a. 6–8 months

    b. up to 2 years

    c. up to 5 years

    d. up to 20 years

14.20 The house mouse is considered to be

a. omnivorous and opportunistic, because they eat all types of food, and anything that is available

b. herbivorous and opportunistic, because they normally only eat plant material but can switch diets if needed

c. opportunistic carnivorous, because their normal diet includes only animal products, but they occasionally feed on plant materials

d. restricted omnivorous, because they eat different types of food, but are very selective about what they consume.

14.21 Unlike rats, house mice can survive without a daily supply of liquid water, because they

a. do not need water to survive

b. can obtain enough moisture from the food they consume

c. breathe enough water to survive

d. can obtain enough moisture from the air using their tongue

14.22 The sense of vision in rats and mice is

a. sharp, light- and dark-adapted, and long-distance ready

b. poor and only accurate during the daytime

c. relatively poor, adapted to nighttime, and color-blind

d. null, because their eyes are not functional

14.23 Feces from Norway rats can be distinguished from those of roof rats because

a. Norway rat feces are smaller (~10 mm), and have pointed ends

b. roof rat feces are larger (~13 mm), and have pointed ends

c. Norway rat feces are larger (~20 mm), and have blunt, rounded ends

d. roof rat feces are smaller (~13 mm), and have blunt, rounded ends

14.24 The number of daily fecal pellets produced by both rats and mice is normally

a. greater than 30

b. less than 20

c. between 10 and 15

d. between 2 and 5

14.25 Rat and mouse runways can be identified in infestations because

a. these animals mark these runways with saliva that dries to a bluish tint

b. these animals repeatedly use these runways leaving them with many food crumbs and feces pellets

c. these animals mark these runways with urine that dries to a greenish tint

d. these animals repeatedly use these runways leaving them clean of dirt or dust but marked with grease marks

14.26 Gnawing damage from rats and mice can be recognized because

a. mice leave incisor teeth groove marks that are 3.5–4 mm wide, while rat marks are 1–2 mm wide

b. mice leave incisor teeth groove marks that are 3–3.5 mm wide, while rat marks are 4–4.5 mm wide

c. mice leave incisor teeth groove marks that are 1–2 mm wide, while rat marks are 2.5–3 mm wide

d. mice leave incisor teeth groove marks that are 1–2 mm wide, while rat marks are 3.5–4 mm wide

14.27 Rodent infestations can be detected even with no visual observation of animals due to

a. a rotten wood odor due to wood decomposition

b. a characteristic odor of rodent urine combined with odor from rodent body glands

c. a tickling sensation due to rodent hairs floating in the air

d. the rainlike noise caused by jumping fleas

14.28 Control of rodents is best accomplished using

a. liquid rodenticide applications

b. a combination of sanitation procedures, rodent proofing, trapping and baiting

c. cats as rodent predators

d. loud sonic devices that scare rodents

14.29 Anticoagulant baits from first and second generations differ in terms of use because

a. first generation anticoagulants only kill older animals (first generation), and the second-generation anticoagulants only kill young individuals (second generation) in rodent populations

b. first generation anticoagulants kill animals that ingest the bait, and the second-generation anticoagulants kill individuals that are born from the generation that fed on the bait

c. first generation anticoagulants require multiple feedings and second generation anticoagulants require only one feeding for a lethal dosage

d. first generation anticoagulants only kill younger animals and second generation anticoagulants only kill older animals in rodent populations

14.30 Non-anticoagulant rodent baits work by

a. either acting as a nerve poison, or by depleting calcium from the bones

b. making the blood very thin so that it cannot circulate properly in the rodent blood vessels

c. making the rodent hungrier so that it consumes the bait until it dies of congestion

d. there is no such thing as a non-anticoagulant rodent bait

15 Birds and Bats

Cities provide abundant resources for birds and bats to colonize. Both birds and bats are not considered pests, but at times their extraction or removal become necessary. Usually, city government and wildlife departments need to be consulted if the problem is severe. Pest managers are capable of handling smaller and localized complaints, but precaution is necessary as these pests and their roosting areas can harbor pathogens and ectoparasites like mites.

Birds

Birds have a higher rate of metabolism which makes them very active during the daytime looking for food. They eat anything from fruit and seeds to insects and worms. Scavenger birds thrive on cooked food and leftovers from human consumption. It is often a subject of debate when their control comes up for discussion. However, birds create conflict when they come into close proximity with humans and human activities. Bird droppings from perches over warehouses, malls and avenue trees can be corrosive on vehicles parked below. Dropping are also a reason for defacement of structures, such as glass windows, and they can carry pathogens. Birds can also be a source of mites. Another common complaint are birds flying across airfields, a serious hazard encountered in many cities around the world.

Control measures must provide a long-term solution to the problem. Many short-term control measures do not work and may pose hazards to both birds and humans. Scarecrows, gun shots, recordings of dying birds, spikes, nets, repellent gels, and other deterrent objects provide temporary relief. Use of birth control baits such as nicarbazin at certain dosages have shown to affect egg laying, interrupting it and reducing hatchability in birds. Other repellent and alarm causing food baits are also available.

© Partho Dhang, Philip Koehler, Roberto Pereira and Daniel D. Dye II 2022. *Key Questions in Urban Pest Management: A Study and Revision Guide* (P. Dhang et al.)
DOI: 10.1079/9781800620179.0015

Bats

Bats often come into conflict with humans as they tend to find shelter in structures. Bats are beneficial animals and participate in pollination of a large number of trees, help in seed dispersal and also consume large numbers of insects. However, their droppings can pose a hazard to the health of humans in close proximity. Bats are also known to act as reservoirs of virus for various zoonotic diseases.

Bat control should be planned with a long-term solution rather than a short-term one. Often short-term solutions can be hazardous to both bats and humans. Solitary bats can enter a building through an open window or door while searching for food and at times large numbers of bats can find shelter in a structure. Thorough inspection is needed to establish all bat exits and entrances and a survey should be conducted to record their pattern of activity.

Bats do not gnaw or claw their way into a building as rats and mice will. Therefore, almost anything can be used as a seal such as fiberglass, wire mesh, plywood, metal, plastic netting or caulking compound.

Key Questions

15.1 Which features are true for pigeons and for sparrows?

    a. Pigeon

    b. Sparrow

    1. Prefers feeding on flat ground

    2. Prefers feeding on grains

    3. Nests are often constructed inside transformers

    4. Often live-in multiple groups in one area

15.2 Pest birds need to be kept away from structures as they

    a. cause food contamination

    b. cause noise pollution

    c. spread disease organisms via contaminated droppings

    d. spread ectoparasites

15.3 _____ is a respiratory disease in humans caused by inhaling spores from the fungus *Histoplasma capsulatum* found in the droppings of birds and bats.

    a. Histoplasmosis

    b. Cryptococcosis

15.4 The house sparrow

    a. does not live in flocks

    b. feeds primarily on grains

    c. makes a nest out of any available material

    d. is primarily a nuisance pest

    e. all of the above

15.5 A bird control program starts with

    a. pre-baiting

    b. installation of exclusion devices

    c. conducting a survey

    d. informing the local authorities

15.6 The following are critical items to be noted when surveying pest birds.

    a. Identify all non-target pest activity on the site

    b. A single site survey is enough for gathering all relevant information

    c. Identify, if dispersed, where the birds would go next

    d. Consider the legal implications

15.7 Nest destruction is effective in keeping pest sparrows in check.

    a. True

    b. False

15.8 Match the following.

    a. Ultrasound devices

    b. Trapping

   c. Bird spikes

   d. Gel

   e. Bird nets

1. Prevents landing on the sticky surface

2. Does not work in controlling birds

3. More effective in controlling pigeons than sparrows

4. Useful in locations where it does not interfere with aesthetics

5. Installed on ledges and prevents roosting

15.9   Bird exclusion devices include the following items.

   a. Electrified wires

   b. Metal sheets

   c. Plastic cover

   d. Mist net

15.10 Pigeons are well controlled by poison baiting, but what needs to be noted?

   a. Bait placement needs to be close to the regular feeding site

   b. Pre-baiting is the single most important aspect of a successful baiting program

   c. Effective baiting will kill the entire flock

   d. Baiting should be done where non-target birds are not present

15.11 The best way to overcome bait shyness is to

   a. increase the amount of bait

   b. suspend baiting for 3 weeks

   c. introduce pre-baiting again after 3 weeks

   d. switch to another method of control

15.12 For effective bird trapping, the following are necessary.

   a. Putting the traps in all visible places

   b. Having a decoy bird in the trap

c. Pre-baiting to check the preferred area of feeding

d. Painting the traps to match the surrounding

15.13 Poison bait such a Avitrol is mixed with untreated corn at the following ratio.

a. 1 part Avitrol to 100 parts corn

b. 1 part Avitrol to 49 parts corn

c. 1 part Avitrol to 29 parts corn

d. 1 part Avitrol to 9 parts corn

15.14 Tips for excluding birds from building areas include

a. broken windows should be sealed

b. signage boards should be placed flat against a building to prevent roosting

c. building ventilators should be netted

d. buildings should be painted with bright colours

15.15 Abandoned and unused parts of structures are susceptible to bat infestation. However, certain structures appear to be more attractive because they

a. are located near the feeding zone

b. maintain ideal temperatures in the roost areas

c. use soft building materials, easily gnawed by bats

d. carry odor from previous bat roosting

15.16 Bats perform the following in our environment.

a. Pollination of fruit trees

b. Removal of insects by eating

c. Dispersal of seeds

d. Produce guano for use as fertilizer

e. All of the above

15.17 Chiropterophily is a term used to describe

   a. a liking for bats

   b. the process of pollination by bats

   c. a phobia of bats

15.18 Birds locate insects by sight, while bats do it by _____ _____.

   a. echolocation

   b. sight and squeaking

   c. using magnetism

   d. using starlight

15.19 Which of these methods can be used best to repel bats?

   a. Bright lights

   b. Naphthalene balls

   c. Ultrasound

   d. Sticky gels

15.20 Two diseases closely associated with bats are

   a. rabies and leptospirosis

   b. leptospirosis and histoplasmosis

   c. histoplasmosis and rabies

15.21 When a single bat finds its way inside a home, the following method may be used.

   a. Use a butterfly net to capture it

   b. Open windows for it to exit

   c. Spray any handy insecticide on hand

   d. Focus bright light on it

16 Pesticides and Formulations

The best practices of pest control utilize a system of urban Integrated Pest Management (IPM) that follows a principle of managing insects and other pests with least risk. However, when prevention fails, pesticides are used to bring a certain level of control.

There are thousands of pesticide products that can be selected for use. There are also many types of pesticide formulations to choose from. Insecticide mode of action is very important for understanding how an insecticide works to kill an insect pest and how to rotate insecticides, so insect resistance is avoided. Below are some of the important groups of insecticides and their modes of action:

- Some of the oldest insecticides still used in urban pest management are the organophosphates (e.g. chlorpyrifos, diazinon) and carbamates (e.g. propoxur).
- Cyclodienes (chlordane, lindane, dieldrin) and fiproles (fipronil) are GABA gated chloride channel blockers.
- Pyrethroids (permethrin, cypermethrin, bifenthrin, deltamethrin, cyfluthrin) are the most frequently used insecticides in urban pest management. These insecticides keep sodium channels open, causing hyper-excitation and nerve blockage.
- Neonicotinoids (imidacloprid, acetamiprid, dinotefuran, clothianidin) affect the receptors for acetylcholine. This causes hyperexcitation of the insect.
- Avermectins (abamectin) change the glutamate-gated chloride channel so they do not accept glutamate, causing paralysis.
- Juvenile hormone analogs (methoprene, hydroprene, pyriproxyfen) mimic the action of juvenile hormones and prevent an immature insect from turning into an adult.

© Partho Dhang, Philip Koehler, Roberto Pereira and Daniel D. Dye II 2022. *Key Questions in Urban Pest Management: A Study and Revision Guide* (P. Dhang *et al.*)
DOI: 10.1079/9781800620179.0016

- Chitin disrupters (benzoylphenyl ureas, diflubenzuron, noviflumuron, hexaflumuron) prevent the formation of a normal insect cuticle after a molt.
- Cellular toxins (boric acid, borax) ingested by insects kill cells in the gut, causing death.
- *Bacillus thuringiensis israeliensis* (Bti) and B. *sphaericus* (Bs) are microbial disrupters of the insect gut, causing death. These are often used for larval mosquito control.
- Electron transport inhibitors (hydramethylnon) affect the ability of insects to have energy for moving or flying.
- Pyrroles and sulfluramid disrupt oxidative phosphoryilation and energy production.
- Hydramethylnon affects electron transport and energy production.
- Sodium channel disrupters (oxadiazines like indoxacarb).
- Ryanodine receptor on muscles (diamides).

It is important to rotate between groups of chemicals every few months in order to prevent or delay the development of resistance.

Insecticides are formulated for use in many ways.

1. Emulsifiable concentrates (EC) are made by dissolving the active ingredient in a solvent with an emulsifying agent. When mixed with water, an emulsifiable concentrate forms an emulsion that is usually a milky white in color.

2. Wettable powders (WP) are made by mixing the active ingredient with a wetting agent to form a dry powder. When the wettable powder is mixed with water, it forms a suspension that requires agitation for uniform mixing.

3. Microencapsulated suspensions or suspension concentrates (ME or SC) are polymer compounds where microcapsules are formed around the active ingredient.

4. Solutions (S) are an active ingredient dissolved in a diluent. Many diluents are oil based. Solutions are usually used as space sprays to kill flying insects. They are often dispensed from pressurized aerosols.

5. Granules are insecticides on an inert carrier like clay.

6. Dusts (D) are dry formulations of active ingredients on a carrier and applied dry.

7. Baits (B) are active ingredients applied to an edible food source for insects. Many are formulated as either granules, gels, or liquids.

8. Fumigants (F) are often gaseous insecticides that must be contained for control.

Key Questions

16.1 What is an insecticide?

    a. A chemical that kills pests

    b. A chemical that kills insect pests

    c. A chemical that contains phosphorus

    d. A chemical based on chrysanthemum flowers

16.2 **What group of insecticides contains phosphorus and oxygen in their core structure?**

    a. Organophosphate

    b. Carbamate

    c. Pyrethroid

    d. Neonicotinoid

16.3 **What group of insecticides has carbon in the center of the molecule with a double bond and a single bond of oxygen?**

    a. Organophosphate

    b. Carbamate

    c. Pyrethroid

    d. Neonicotinoid

16.4 **Both organophosphates and carbamates kill insects in a similar way. What is the mode of action for these two groups of chemicals?**

    a. Disrupts axial nerve function

    b. Activates nicotinic acetylcholine receptor

    c. Acetylcholinesterase inhibitor

    d. Affects GABA and glutamate gated-chloride channels

16.5 **What group of insecticides has a triangular-shaped cyclopropyl group in the molecule and is derived from chrysanthemic acid?**

    a. Organophosphate

    b. Carbamate

    c. Pyrethroid

    d. Neonicotinoid

16.6 **How does a pyrethroid insecticide affect the insect's nervous system?**

    a. Disrupts axial nerve function

    b. Activates nicotinic acetylcholine receptors

    c. Acts as an acetylcholinesterase inhibitor

    d. Affects GABA and glutamate-gated chloride channels

16.7 **What group of insecticides is based on nicotine?**

    a. Organophosphate

    b. Carbamate

    c. Pyrethroid

    d. Neonicotinoid

16.8 **Select the name of the chemical that is not a neonicotinoid.**

    a. Imidacloprid

    b. Acetamiprid

    c. Bifenthrin

    d. Thiamethoxam

16.9 **Select the name of the chemical that is not a pyrethroid.**

    a. Permethrin

    b. Cypermethrin

    c. Bifenthrin

    d. Thiamethoxam

16.10 **Select the name of the chemical that is not an organophosphate.**

    a. Propoxur (Baygon)

    b. Chlorpyrifos

    c. Diazinon

    d. Dichlorvos

16.11 **Select the name of the chemical that is a carbamate.**

    a. Chlorpyrifos

    b. Propoxur (Baygon)

c. Diazinon

d. Dichlorvos

16.12 Choose the insecticide that limits energy production in insects causing them to become slow and uncoordinated.

a. Fipronil

b. Cypermethrin

c. Imidacloprid

d. Hydramethylnon

16.13 If you were using bifenthrin for years to control a pest and it became resistant, which chemical listed below would be a good choice to use for the next treatment?

a. Cypermethrin

b. Permethrin

c. Imidacloprid

d. Cyfluthrin

16.14 An example of a cellular toxin is

a. boric acid

b. malathion

c. fipronil

d. abamectin

16.15 Chemical control is not the only method for providing pest management. Which of these options is a method of physical control?

a. Ultrasound devices

b. Vacuuming and physically removing pests

c. Use of natural enemies

d. Fumigation

16.16 Which formulation of insecticide would be a poor choice for treating concrete blocks or slabs?

a. Emulsifiable concentrate

b. Wettable powder

c. Granules

d. Suspension concentrate

16.17 Which type of control has been shown to be ineffective or not appropriate for bed bug control, and possibly German cockroach control?

a. Residual sprays to cracks and crevices

b. Bed bug sorptive dusts

c. Boric acid dust

d. Total release aerosol bombs

16.18 Which insecticide formulation should not be applied under or near refrigerators or appliances with fans?

a. Emulsions

b. Suspensions

c. Gel baits

d. Dusts

16.19 What is a hazard of total release insecticide bombs that has nothing to do with insecticides?

a. Inhalation

b. Contamination of surfaces

c. Explosion

d. Spillage

16.20 What equipment is needed for a thorough home pest inspection?

a. Flashlight

b. Mirror

c. Screwdriver

d. Paper and pencil

e. All of these

16.21 **Making a connection with your customer is very important in establishing an effective pest management program. What is the first step in making that connection?**

a. A smile and confident eye contact

b. A professional greeting

c. Sincere compliments

d. An appropriate handshake

16.22 **Juvenile hormone analogs are insect growth regulators that are used for which type of pests?**

a. Occasional invaders

b. Cockroaches, fleas, and mosquitoes

c. Bed bugs

d. Termites

16.23 **Chitin synthesis inhibitors are insect growth regulators that prevent an insect from developing a new exoskeleton. What are chitin inhibitors most frequently used to control in urban pest management?**

a. Termites

b. Ants

c. Flies

d. Cockroaches

16.24 **Insecticide resistance is the development of the ability to survive insecticide doses that were previously toxic. What is a way to avoid the development of insecticide resistance?**

a. Use different formulations of an insecticide

b. Use different methods of treatment

c. Rotate to different insecticide groups

d. Use different manufacturers of an active ingredient

16.25 **Which group of insecticides affects insect muscles causing them to not respond to nervous stimuli?**

a. Diamides

b. Oxadiazines

c. Pyrethroids

d. Neonicotinoids

16.26 **It is important to perform effective pest management with minimal risk. What is a method of reducing risk while still controlling insects?**

a. Spray on a calendar basis – whether pests are present or not

b. Apply insecticides to areas where children and pets have contact

c. Apply insecticides when pests are present and to their hidden harborages

d. Not allowing insecticides to dry before allowing customer reentry

16.27 **What is a method of preventing inhalation of insecticides that often appears on the product label?**

a. Hold your breath

b. Use a particle respirator

c. Wear a bandana

d. Wear shoes

16.28 **What are some appropriate methods for monitoring pests?**

a. Looking for pest infestations

b. Placing sticky traps

c. Flushing and counting pests

d. All of the above

16.29 **What is the first step in handling a pesticide spill?**

a. Control the spill

b. Read the label

c. Contain the spill

d. Clean up the spill

16.30 Where should pesticides be stored on a service vehicle?

a. In the bed of a truck

b. Next to the spray tank so they are handy for dilution

c. In the passenger compartment of a locked service vehicle

d. In a locked lock box

17 Handling Pesticides

Pesticides are one of the most important and beneficial tools for the urban pest management industry. However, pesticides are toxicants that affect most organisms, including humans. The pesticides that are used in urban areas are selective in their toxicity to affect the survival, development, and reproduction of the target pest species. These pesticides have greater adverse effects on pests than on humans and other non-target species. However, applicators should not underestimate the effects of pesticide exposure when they use improper mixing, application, and safety equipment, and do not take precautions to protect themselves. Pesticide safety awareness can minimize the potential adverse effects of excessive exposure on the job.

Toxicology is the study of the adverse effects of pesticides. The basic principle of toxicology is the concept of dose. Dose is the amount of toxicant per weight of the exposed organism. The concept of dose means that insecticides are inherently more toxic to insects than people.

One of the methods of comparing the hazards of pesticides is to determine a lethal dose 50 (LD-50; the amount of toxicant to that would kill 50% of the exposed organisms). The LD-50 is expressed as mg of toxicant per kg of body weight.

The route of pesticide exposure is important in determining toxicity. The typical routes of exposure are oral (by mouth), dermal (exposure to the skin), ocular (exposure through the eyes), and inhalation (by breathing airborne insecticides). Most exposure by pest control applicators is through the skin, and various parts of the body have different absorption properties. For instance, compared to the forearm, about 4 times more chemical is absorbed through the forehead and scalp and 11 times more in the genital area. Because commercial pesticide applicators have continuous exposure to insecticides, it is important that appropriate personal protective equipment (PPE)

© Partho Dhang, Philip Koehler, Roberto Pereira and Daniel D. Dye II 2022. *Key Questions in Urban Pest Management: A Study and Revision Guide* (P. Dhang et al.)
DOI: 10.1079/9781800620179.0017

be used as directed on the pesticide label. Personal protective equipment is usually recommended to protect the body from skin absorption, the respiratory system from inhaling sprays and dusts, the eyes to protect eyes from irritating or toxic chemicals, and hand and feet protection when handling pesticides or walking through treated areas.

It is necessary for an applicator to be familiar with various hazard symbols indicated in the labels as well as with the signs and symptoms of excessive pesticide exposure and a knowledge of how to respond.

Key Questions

17.1   What is the way toxicity is measured and usually expressed?

    a. LD-20

    b. LD-40

    c. LD-50

    d. LD-100

17.2   An LD-50 of less than 50 is defined as a _____ pesticide.

    a. highly toxic

    b. moderately toxic

    c. slightly toxic

    d. practically no toxicity

17.3   Toxicity is a function of

    a. exposure to a toxic substance

    b. dose multiplied by time

    c. LD-50s

    d. mode of action

17.4   What is a dose called that would not produce an adverse effect during the life of a species?

    a. LD-100

    b. LD-50

    c. NOEL

    d. Practically no toxicity level

17.5 **What is the term for the production of benign or malignant tumors (cancer) by a chemical?**

a. Carcenogenesis

b. Acute toxicity

c. Chronic toxicity

d. Oncogenesis

17.6 **The most common route of entry for pesticides into children is**

a. ocular

b. oral

c. inhalation

d. dermal

17.7 **An applicator mixing and loading pesticides without gloves would probably have the most exposure due to _____ exposure.**

a. ocular

b. oral

c. inhalation

d. dermal

17.8 **An applicator fogging for stored product pests should use a respirator to prevent _____ exposure.**

a. ocular

b. oral

c. inhalation

d. dermal

17.9 An applicator measuring out an emulsifiable concentrate at eye level should wear eye protection to prevent _____ _____ exposure.

    a. ocular

    b. oral

    c. inhalation

    d. dermal

17.10 Eating or smoking without washing hands after exposure to pesticides can result in _____ exposure.

    a. ocular

    b. oral

    c. inhalation

    d. dermal

17.11 Every action has risks. Which has the greatest risk for people?

    a. Pesticides

    b. Alcohol

    c. Smoking

    d. Motor vehicles

17.12 LD-50 is expressed as

    a. mg/kg

    b. lb/ton

    c. g/kg

    d. ounces/gallon

17.13 What is the Federal Agency that regulates insecticides in the US?

    a. USDA

    b. OSHA

    c. DOD

    d. EPA

17.14 What is the federal agency that provides regulations to protect workers on the job?

    a. USDA

    b. OSHA

    c. DOD

    d. EPA

17.15 What is the best way to learn how to use a pesticide product?

    a. SDS

    b. MSDS

    c. The product label

    d. Internet

17.16 What is the best way to learn about how to clean up a pesticide spill?

    a. SDS

    b. MSDS

    c. The product label

    d. Internet

17.17 What is the term for the images on an SDS that give an indication of the type of hazard associated with a product?

    a. Animations

    b. Imogees

    c. Pictograms

    d. Pictures

17.18 Clothing contaminated with pesticides needs to be

    a. put in the wash with family clothes

    b. washed with only bleach to decontaminate the pesticides

    c. washed separately from other laundry using hot water and detergent

    d. discarded

17.19 Pesticide storage should be labeled with which level of signage?

    a. Danger

    b. Warning

    c. Caution

    d. The signal word for the most hazardous product

17.20 Drains in the pesticide storage area should

    a. be attached to the sewer system

    b. not be there. The storage area should contain spills

    c. should drain to a ditch outside

    d. have an absorbent material to make removal easy

17.21 If a pesticide label says an applicator must wear long sleeved shirt, long pants, and shoes plus socks, which choice would be appropriate protection?

    a. Disposable sleeves

    b. Work shoes

    c. Athletic shoes

    d. Golf shirt

17.22 What is an appropriate glove material for an "E" rated pesticide solvent protective material on the pesticide label?

    a. Polyethylene

    b. Nitrile rubber

    c. Natural rubber

    d. Polyvinyl chloride

17.23 **A pesticide label states "all pesticide handlers must wear protective eyewear (goggles and/or faceshield and/or shielded safety glasses with front, brow and temple protection)". Which safety equipment is appropriate?**

a. Sunglasses

b. Corrective lens glasses

c. Contacts

d. Safety glasses with brow and temple protection

17.24 **What respirator that gives 95% efficiency, should be used when applying an oil-based insecticide aerosol? See chart below:**

Filter efficiency	N – Not resistant to oil	R – Partially resistant to oil	P – Strongly resistant to oil
95%	N95	R95	P95
99%	N99	R99	P99
100% (99.97%)	N100	R100	P100

a. N99

b. N95

c. P100

d. P95

17.25 **There is a difference between respirator filters that remove particles and vapors. A particulate filter is often color coded. Which color denotes a particle filter?**

a. Black

b. Red

c. White

d. Magenta

17.26 Which color vapor filter should be selected for most insecticides?

    a. Black

    b. Red

    c. White

    d. Magenta

17.27 What is the best color combination for filters for an aerosol insecticide application?

    a. Black and white

    b. Magenta and black

    c. Red and white

    d. White and magenta

17.28 Is it against the law to use the wrong safety equipment listed on the label?

    a. Yes

    b. No

17.29 What is the law that regulates pesticides in the US?

    a. FEPCA

    b. OSHA

    c. DOD

    d. FIFRA

17.30 What agencies implement the provisions of the Federal Insecticide, Fungicide and Rodenticide Act?

    a. USDA

    b. State Departments of Agriculture

    c. EPA

    d. State Departments of Environmental Regulation

18 Integrated Pest Management

Integrated Pest Management (IPM) is a concept first designed for agriculture in response to the increase in the usage of chemical pesticides. Today, IPM is a methodology practiced by urban pest managers which employ human judgment in their act. The reason for the necessary shift to IPM from conventional pest management activity was because over reliance on pesticide had led to repeated failures. In addition, the discovery of the harmful effects of pesticides soon became evident, and developing an alternative strategy was inevitable. The demonstrated successes of IPM methodologies have helped expand the subject to other specialized area of pest management, such as Integrated Vector Management (IVM) and Integrated Termite Management (ITM).

Conventional methods of pest control which use pesticide as a "stand-alone" tool can cover a wider range of pests, provide quick and easy elimination and have long field persistence as key benefits. Pesticides provides significant or acceptable reductions in pest populations immediately. The process involves a single action of chemical application following a regular, predetermined spray schedule. However, IPM is about more than eliminating pests. Maintaining control over pests and preventing re-infestations is given more importance. IPM programs have a number of key attributes to offer when it comes to maintaining control. Each IPM program follows stringent monitoring and intervention methods to keep check on pest population. This usually involves a combination of numerous methods such as non-chemical tools, barriers, education and correct pesticide use.

An IPM program is dependent on training and requires skill and knowledge. IPM is dependent on professionals who are best at keen observation, finding the source, analyzing each situation, determining the pest threshold, identifying pest exclusion methods and are able to quickly assimilate and implement emerging research. Thus, pest managers trained in conventional

© Partho Dhang, Philip Koehler, Roberto Pereira and Daniel D. Dye II 2022. *Key Questions in Urban Pest Management: A Study and Revision Guide* (P. Dhang *et al.*) DOI: 10.1079/9781800620179.0018

pest control often find it difficult in adapting to the multi-tasking protocols required in implementing IPM.

It has been reported that issues such as psychological resistance to change, loss of authority, resistance in learning new technologies, general fear of failure, fear that IPM will restrict use and access to pesticides, and that IPM is more expensive than conventional pest control are among reasons for the poor adoption of IPM.

Key Questions

18.1   What does the acronym IPM stand for?

    a. Intelligent Pest Management

    b. Integrated Pest Management

    c. International Pest Management

18.2   Integrated Vector Management (IVM) is a World Health Organization led program to control

    a. mosquitoes

    b. mosquitoes, flies, ticks, bugs and other vectors

    c. all neglected tropical diseases

18.3   The single best way to define IPM is

    a. using a number of tools such as traps, monitors and pesticides together to perform a job

    b. a method where pesticide use is totally removed

    c. a process of integrating education, judgment, reasoning and the correct products in a job

    d. to involve entomologists in performing a job

18.4   IPM in structures can be achieved by keeping a check on the following:

    a. housekeeping

    b. maintenance

    c. behavior of residents

    d. building materials used

    e. all of the above

18.5 **A consultation between a builder/architect and an integrated pest management expert is considered important in establishing a pest-free structure.**

    a. True

    b. False

18.6 **The following criteria are considered part of facility designing, helpful in keeping pests away.**

    a. The type of lighting and solid waste management

    b. The quality of building materials used

    c. Design of the landscape

    d. The color of the paint used on the outer wall of the structure

18.7 **Sodium vapor lights are less attractive to insects than other lights.**

    a. True

    b. False

18.8 **IPM for homeowners can be made cost effective by**

    a. tailoring the requirement to each property and situation

    b. choosing the cheapest chemicals

    c. doing a one time whole house treatment

    d. signing up for a regular maintenance service

18.9 **Why was a need for IPM felt in the pest management industry?**

    a. To increase reports on pest control failures

    b. In an effort to reduce pesticide dependence

    c. To reduce pollution caused by pesticides

    d. In order to promote non-chemical products

18.10 **Rachel Carson's 1962 book *Silent Spring* is known for the following:**

    a. The book became a textbook on pest management

    b. The book documented the effects of chemicals on non-target organisms

    c. The book shifted public opinion towards safer methods of pest control

    d. The book eventually led to the setting up of the EPA in the USA

18.11 The gravest concerns of using pesticide in close proximity comes from the proven fact that they have the highest risk to

    a. pets

    b. pediatric population

    c. pregnant women

    d. food products

18.12 One of the best methods to prevent pest infestation in a home is by

    a. sealing all crack and crevices

    b. advising residents on possible materials that could carry pests

    c. checking on the sanitation of the yard

    d. having a monthly maintenance service for pest control

18.13 Record keeping in an IPM program is used to keep a periodic check on

    a. the number of people going in and out of a site

    b. structural deficiencies

    c. pest sightings

    d. observations of the housekeeper

    e. the weather

18.14 Communication between property owners and the IPM managers is not considered a critical criterion in effective running of a program.

    a. True

    b. False

18.15 What is the major route of pesticide exposure to an indoor population?

    a. Indoor spraying

    b. Outdoor spraying

c. Off-the-shelf consumer products

d. Food

e. all of the above

18.16 Public use of pesticide is on the increase due to the following:

a. awareness that pests are injurious

b. increase in quality of living standards

c. intolerance to foreign organisms

d. advertisement

e. all of the above

18.17 What is the single most important drawback in application of a single pesticide repeatedly at a site?

a. The customer is not impressed

b. The insect acquires resistance to the chemical

c. The product loses its efficacy

d. The environment gets contaminated

18.18 Food-based baits are commercially available to effectively control

a. termites

b. stored product pests

c. bed bugs

d. rodents

18.19 Food-based insecticide baits can be used as a standalone product in IPM programs because these products

a. are available to consumers in stores

b. are easily handled without skill

c. can eradicate a population completely

d. contain fewer active ingredients

18.20 The right person to determine pest threshold in a structure would be the

    a. pest control technician

    b. owner of the pest control company

    c. customer

    d. city health officer

18.21 In an infested structure, pests are evenly distributed.

    a. True

    b. False

18.22 In general terms, a conventional pest management program means

    a. use of a stand-alone methodology

    b. use of both baiting and spraying together

    c. use of the latest registered product from the market

    d. total eradication of the pest

18.23 Knowledge of pest biology and ecology is a must in implementation of IPM because

    a. it provides information on population dynamics of the pest

    b. it provides critical control points on the pest

    c. it helps to keep the customer updated

    d. it helps to justify the higher cost

18.24 Success of integrated pest control is more visible in agriculture than in urban/household pest control.

    a. True

    b. False

    c. No such evidence available

18.25 Arrange the following in the right sequence when undertaking IPM.

    a. A written report

    b. Pesticide application

c. Sealing the gaps

d. Inspection

18.26 Which among the following is to be avoided when implementing an IPM program?

a. Use of multiple pest control methods throughout the program

b. Inspection, sampling and analysis to determine any control method

c. Providing a continuous education program to the customer

d. Use of a pre-established schedule of work established elsewhere

18.27 Match the following for the best action.

a. Integrated pest management

b. Preventive pest management

c. Reactive pest management

d. Eradication program

1. Makes use of the fastest method to eliminate the pest population without any regards to acceptable threshold

2. Makes use of monitoring and traps in the action plan

3. Makes use of an elaborate inspection and recommendations to bring pest population to an acceptable threshold level

4. Makes use of whatever is available to control the situation

18.28 Poor adoption of IPM in certain segments is due to

a. poor success in using IPM

b. fear of not being allowed to use chemicals

c. it being perceived as expensive

d. the fact that all pests cannot be covered by IPM methods

18.29 An IPM program designed for German cockroaches includes the following:

a. surface residual spray and a report

b. baiting, sanitation, trapping and exclusion

c. surface spray, baiting and trapping

19 Answers

1 Introduction to Urban Pest Management

1.1 b	Humans have modified their environment and are intolerant of other organisms occupying the same space as them and adversely affecting their objectives.
1.2 a and b	A small number of organisms have associated very closely to humans and their environment. These organisms predominantly come into contact with human and human-made objects mostly in urban areas.
1.3 c	Urban pests constitute both invertebrates and vertebrates. At times invasive plants may also be termed as pests.
1.4 a	Human activities such as food preparation and storage, storing items, creating clutter, poor lighting and ventilation, etc. create an ideal environment for pests to be attracted into the structure, eventually making them thrive.
1.5 a	Globally, humans continue to encounter major threats from arthropod-borne diseases, such as malaria, dengue, dengue hemorrhagic fever, Japanese encephalitis and plague.
1.6 a and c	Delusional parasitosis is a psychiatric condition where people have the mistaken belief that they are parasitized by bugs, worms, or other creatures.
1.7 a-2; b-4; c-1; d-3	
1.8 a and c	Body parts and the waste of dust mites are human allergens. Whereas the saliva, feces and shedded body parts of cockroaches can trigger both asthma and allergies.

© Partho Dhang, Philip Koehler, Roberto Pereira and Daniel D. Dye II 2022. *Key Questions in Urban Pest Management: A Study and Revision Guide* (P. Dhang *et al.*)
DOI: 10.1079/9781800620179.0019

1.9 a, b, and c	At times, cockroach allergy may trigger asthma, and this will lead to difficulty breathing, chest tightness or pain, and a whistling or wheezing sound when breathing out.
1.10 a, b, and d	Many species of the phylum Arthropoda, which includes insects, arachnids and crustaceans are capable of exuding toxins that may cause serious reactions in the human body.
1.11 a-2; b-5; c-4; d-1; e-6; f-3	
1.12 b	The insect vectors responsible for the spread of human disease include Diptera (mosquitoes and biting flies), Hemiptera (true bugs), Anoplura (lice) and Siphonaptera (fleas).
1.13 a, b and c	Generally, reports of bites or stings, or reaction from bites and stings come much later after the injury has actually been inflicted. In this scenario, sealing the area does not serve any purpose.
1.14 b and c	Pests look for shelter and food in any structure, and to deny these is the best way to deter them in long run.
1.15 a, b, and c	Pest threshold is the pest density over which a control tactic must be implemented to avoid loss. Certain sites may require a "zero" pest tolerance and others may operate on a flexible threshold limit.
1.16 a-2; b-1; c-4; d-3	
1.17 a and c	Until the 1940s, inorganic compounds and botanicals providing broad spectrum pesticidal properties served as pest control products, which are collectively known as "first generation" pesticides.
1.18 a	All types of development modifiers such as moulting hormones, growth inhibitors and chitin synthesis inhibitors are classified as third generation pesticides.
1.19 b and c	1940–1970 was the period of introduction of synthetic organic molecules belonging to organochlorines, organophosphates, carbamates and pyrethroids groups. These are all classified as second generation pesticides.
1.20 e	Chitin is the most abundant natural amino-polysaccharide and serves as a structural component of fungus, spores, cell walls, and is present in arthropod cuticles, etc. Plants and vertebrates including humans do not produce chitin.
1.21 b, c and d	DDT caused damage to organisms, and some animals exposed to DDT in studies even developed liver tumors. As a result, today, DDT is classified as a probable human carcinogen.
1.22 a, b and d	A pheromone is a secreted chemical that triggers a social response in members of the same species. The social response could be varied, such as attraction, aggregation or alarm.

1.23 a	The first active pheromone was chemically identified in 1959 from the silkworm moth, *Bombyx mori*, and was named bombykol.
1.24 b	In modern day pest control, pesticide is considered as the last tool of use. Its use becomes inevitable only when other methods fail to bring the desired result.
1.25 b	
1.26 a-3; b-4; c-1; d-2	
1.27 a	In practical terms, insects that repeatedly fail to be adequately controlled by the label rate of an insecticide are said to be insecticide resistant. This mostly happens when the same chemical is used over a long period.
1.28 e	Insecticide resistance can be divided into behavioral resistance and physiological resistance. In behavioral resistance, the insect populations develop the ability to avoid insecticide exposure. Whereas, in physiological resistance the insect undergoes modification such as increased metabolic detoxification, and decreased target site sensitivity to overcome the pesticide.
1.29 b, c and d	Responsible pest management is the correct definition of "green" pest control.
1.30 a	

2   Pest Identification

2.1 b	This species has a bright silvery lyre-shaped dorsal pattern and white banded legs. As its name implies, this mosquito is capable of transmitting yellow fever, as well as dengue, Chikungunya virus, Zika virus and dog heartworm.
2.2 a	This species has a single longitudinal silvery dorsal stripe and white banded legs. Now found worldwide, it has expanded its range into Europe, North and South America and Africa. The asian tiger mosquito is known to transmit several diseases including West Nile virus, dengue virus and eastern equine encephalitis virus.
2.3 c	This mosquito has the unofficial name of white-footed woods mosquito. They are native to North and South America and found mainly in woodland habitat. It has an aggressive feeding nature with a painful bite to humans. Although some viruses have been isolated from this species, they are not considered to be of major concern to humans. However, there have been cases of Venezuelan equine encephalomyelitis virus in humans reported in Central and South America.

2.4 a-2; b-3; c-1	When mosquito eggs hatch, larvae known as wigglers emerge and go through four larval instars. The larvae breathe air through spiracles and obtain air at the water surface. Anopheline larvae have spiracular openings along most abdominal segments. The spiracles are surrounded with palmate hairs that keep them in touch with the water surface. *Culex* and *Aedes* larvae have a pair of spiracles that open onto a respiratory siphon which is an elongate respiratory tube, differing in sizes.
2.5 b	Note, the common bed bug, on the right, has a more excavated, or u-shaped pronotum than the tropical bed bug, on the left.
2.6 c	House flies are about 3 to 6 mm long. They are dull gray in color and have four narrow black longitudinal stripes on their thorax. House flies feed on a wide variety of things. They are attracted to homes and other structures by odors and air currents. On cool days they are attracted to the warm air currents produced by structures while on warm days the cooler air currents escaping from windows and doors attract them.
2.7 b	They are attracted to the early stage decaying and fermenting of fruits and vegetables while others of this genus may be attracted to human and animal excrement. Because of their varied feeding habits, it is believed they may be vectors of disease. This genus contains a rather large number of species. *Drosophila melanogaster* is the species most often found in homes and other buildings such as grocery stores, restaurant kitchens and fruit packing plants.

2.8 d	Notice the unique pattern in the wings. The moth-like appearance is caused by all the tiny hairs that cover the wings and body. They are known to breed primarily in drains, which is where they may contract disease-causing pathogens and bacteria. This could be a serious problem for health care and food handling facilities.
2.9 a	Note the ootheca still attached. The average size for an adult is about 2" in length and is the largest of the peridomestic roaches most commonly found in North America. She will deposit her egg capsule (ootheca) in a crack or crevice within four days after it is formed. On average, she will produce 16 ootheca in her lifetime with each one containing 16 eggs. Because this roach is known to travel within sewer systems from building to building, they can become a public health problem.
2.10 a	Its origin is thought to be Africa and was first discovered in the United States in 1903. Some research suggest it was introduced to Europe in the late 1930s and early 1940s during the Second World War by US Military troops. At one time they were commonly found in structures, but now are not as common, possibly because of home air conditioning which reduces the overall temperatures they require during the warmer seasons.
2.11 c	It has a world-wide distribution and is the most common species found to infest homes and other man-made structures. They are synanthropic, meaning strictly associated with humans. The female will produce, on average, up to nine ootheca containing 30–48 eggs within. She will carry the ootheca until just before they hatch and place it in a secure place. Just as other pest cockroaches, they are known to carry pathogens and allergens that could be harmful to humans.
2.12 d	Its distribution is worldwide, but believed to have originated from Southern Russia. They are a rather large roach, measuring up to approximately 29 mm (1 inch). Neither the male or female can fly, although the male does have small wings.
2.13 c	The key to identifying the German cockroach (*Blattella germanica*) ootheca is the shape and size. It's less than a ¼" long with about 16 subsegments being very apparent. The ootheca will contain 30–48 eggs within.
2.14 a	The key to identifying the American cockroach (*Periplaneta Americana*) ootheca is its shape and size. It is more than a ¼" long, symmetrical and the length is less than twice the width. It contains 16 eggs total, 8 per side.

2.15 b	The origin of this species is believed to be from Taiwan and Southeastern China. It is now found in several locations on the globe including South Africa, Japan, and North America. It has the widest distribution of all from the genus *Coptotermes* and is considered the most important economically. Notable is the oval or tear drop shape of the head. By contrast soldiers in the genus Reticulitermes have a rectangular shaped head.
2.16 a	*P. megacephala* is capable of becoming highly invasive and is known to displace native ant species. It is thought to be native of Africa because of its extensive geographic coverage there. Populations are found throughout the world in subtropical and tropical regions. The workers are dimorphic (major and minor workers).
2.17 d	Note the other possible answers listed are all two-node ants. The Argentine ant is a one node ant. Many pest ants can be identified by using this key as a starting point.
2.18 a	The brown dog tick is found throughout the world. Its primary host is dogs, however, they may feed on other animals including humans. Unlike other tick species, the brown dog tick can complete its life cycle outside and indoors. They have been known to be vectors of several disease pathogens in dogs. The brown dog tick has also been identified as a reservoir of *Rickettsia rickettsii*, which causes Rocky Mountain spotted fever.
2.19 a	Silverfish are one of the most primitive insects on earth, dating back to over 400 million years ago. They feed on almost anything in our homes. Book bindings, hair, clothing, carpet, glue, coffee, sugar, paper, leather, cereals and dandruff. Silverfish can live for 12 months or more without feeding, as long as water is available.
2.20 c	They are the largest eusocial wasp native to Europe. Although a non-native, it is the only true hornet (*Vespa*) found in North America. Its sting is likened to that of a honey bee despite its large size. They are known to be nocturnal and are often attracted to lights at night.
2.21 a	Shown in the photo is a North American species, *Priobium sericeum*. They are known to infest flooring and understructure wood members of structures in the eastern United States of America.

2.22 a-3; b-1; c-2	House mouse droppings are 3–6 mm long, rod shaped with pointed ends and lack ridges; roof rats droppings are 12.5 mm long, spindle shaped with pointed ends; Norway or sewer rat droppings are 20 mm long, capsule-shaped with blunt ends.

3   Mosquitoes

3.1 c	The family name for mosquitoes is Culicidae. All the Culicidae have three pairs of long, thin legs, long mouthparts (proboscis), and long filamentous antennae. The wings have scales.
3.2 b	Mosquito larvae are aquatic. They breathe on the water surface through their siphon or spiracles surrounded by water-repellent hairs. All mosquito larvae require water sources to survive.
3.3 a	Female mosquitoes feed on blood. The males do not feed on blood. However, both male and female mosquitoes will feed on nectar for carbohydrates.
3.4 b	*Aedes aegypti* and *Aedes albopictus* usually have their peak feeding activity during daytime hours. Most mosquitoes are active when the humidity is high and winds have subsided.
3.5 a	Male mosquitoes have antennae that are densely covered with long hairs. Those antennae are called plumose. The hairs are used by male mosquitoes to hear female mosquitoes flying.
3.6 e	Most mosquitoes have a siphon tube for obtaining air at the water surface. But Anopheles mosquitoes have the spiracles on the abdominal segments surrounded by palmate hairs that are water repellent.
3.7 b	Culex mosquitoes lay their eggs in rafts. The eggs are elongate and are tied together along their sides. These rafts are laid and float on the water surface. They can have from 100–400 eggs. Each egg has a float to keep it from sinking.
3.8 a	The larvae of mosquitoes are filter feeders and consume algae, bacteria, protozoans, and small pieces of organic matter. The food is drawn into their mouths by mouth brushes.
3.9 c	Most male mosquitoes live only 6–7 days after emergence. They need to find a mate almost as soon as their adult cuticle hardens, and wings expand. Males from a brood emerge a little faster than the females.
3.10 b	Females take a blood meal in order to have the nutrition to develop eggs. It usually takes 3 days for the mosquito to digest the blood and for the eggs to be developed and fertilized.

3.11 d	Female mosquitoes are able to filter the red blood cells from excess water. Those cells contain most of the protein and nutrients for egg development. The excess water is excreted and that process of eliminating excess water is called pre-diuresis.
3.12 c	*Aedes albopictus* is the Asian tiger mosquito and is considered a major urban pest in many parts of the world. It is capable of transmitting Zika, dengue, and yellow fever virus and other disease organisms as well. This mosquito breeds in small containers and small bodies of water.
3.13 a	*Culex pipiens* (*quinquefasciatus*) is called the common house mosquito. It breeds in small containers, but lays its eggs in rafts on the water surface. This species is one of the main vectors of West Nile virus.
3.14 b	The common malaria mosquito is *Anopheles quadrimaculatus*. It gets its species name from the four dark spots near the center of its wings. This mosquito develops in bodies of fresh water containing aquatic or marsh vegetation.
3.15 d	*Aedes taeniorhynchus* is the black salt marsh mosquito. These mosquitoes lay their eggs on the edges of brackish water. When there are high tides or floods, these eggs will hatch. The eggs can survive desiccation for about 5 years.
3.16 c	Anopheles mosquitoes transmit malaria. The most effective program for eliminating malaria was to prevent the mosquito vectors from entering buildings and biting sleeping or resting people. Elimination of this mosquito from the US was achieved by advocating window screens.
3.17 a	Piles of discarded tires are great places for the development of large numbers of either *Aedes aegypti* or *Aedes albopictus* mosquitoes. In fact, the introduction of *Aedes albopictus* into the US was traced to old tire shipments from northern Asia in the 1970s.
3.18 c	Because mosquitoes breath through a siphon, the application of oil on the water surface will suffocate them when they try to breathe. Monomolecular films and surface tension reducers effectively suffocate mosquito larvae.
3.19 a	Bti is a biological larvicide for mosquitoes. It is usually applied as briquets that float on the water surface. The toxin is ingested by feeding mosquito larvae and must remain on the water surface to be effective.
3.20 c	Most insect growth regulators (IGRs) specifically affect insects. When late-stage mosquito larvae are exposed to IGRs, they do not molt successfully into adults. They usually die in the pupal stage as deformed adults.

3.21 a	Methoprene is an IGR that is a juvenile hormone mimic that is widely used for larval mosquito control, even though it kills mosquitoes in the pupal stage. Bti is a biologically produced product that kills mosquito larvae when they ingest it. Naled and malathion are traditional insecticides used for adult mosquito control.
3.22 b	ULV applications are used to kill flying mosquito adults. Very small amounts of insecticide are used and are in small droplets that stay airborne for long periods of time. They impinge on flying mosquitoes, effectively killing them.
3.23 d	The optimal size for a ULV droplet would be about 20–30 microns in size. That size droplet will impinge on the mosquito wings and body providing optimal kill.
3.24 a	Mist blowers are used to kill mosquitoes that land and rest on vegetation and structures. Mosquitoes do not fly all the time. A mist blower puts out a residual spray that effectively treats vegetation.
3.25 c	Biological control is the use of predators, parasites and microorganisms to control pests. Two successful uses of biological control are mosquitofish, like Gambusia, and other top feeding minnows, and also the use of copepods.
3.26 b	Repellent is an essential item of PPE that would protect an applicator from mosquito bites and mosquito transmitted disease. When dealing with mosquitoes a repellent is a very important piece of PPE.
3.27 c	Diethyl toluamide (DEET) was developed to protect troops from disease carrying mosquitoes during the Second World War. Some formulations of DEET are up to 90% active ingredient, but concentrations above 35% do not improve efficacy.
3.28 a	Octenol was synthesized from cattle breath and is often used to improve the catch of mosquitoes in light traps used for mosquito surveillance. Other attractants used with it might be carbon dioxide and lactic acid. The scents associated with breathing and sweating are the cues for mosquitoes to find a host and get a blood meal.
3.29 b	An ovitrap is used to capture egg laying mosquitoes for either monitoring or control. Ovitraps capture these females when they enter the device to deposit eggs onto or next to water in the trap.
3.30 d	Mosquito larvae are usually on top of the water where they feed and breathe. A dipper is often used to skim the surface of the water to collect larvae.

4   Bed Bugs

4.1 b	Comparison of the pronotum are easier when one has available specimens from both species. Infestations of the tropical bed bug are more prevalent in warmer areas of the globe, but the common bed bug has spread into areas previously dominated by the tropical bed bug.
4.2 c	Bed bugs are in the stink bug order Hemiptera that has sucking mouthparts, and in the family Cimicidae, that contains many blood-sucking temporary ectoparasites. The genus Cimex includes species that have been associated with humans through recorded history.
4.3 d	The common bed bug was originally associated with bats in the Old World and is considered a cosmopolitan species, whereas the tropical bed bug is more restricted to tropical areas.
4.4 d	Besides humans, bed bugs are known to utilize bats and chickens as hosts, although they have been observed feeding in many other animals.
4.5 a	The only food that bed bugs ingest is blood, which can be obtained from different hosts, although the main hosts are humans and bats.
4.6 a	Bed bugs can survive for long periods without feeding, but they require blood meals to develop and produce eggs.
4.7 c	Bed bugs are small, but large enough in the late larvae stages and adult stage to be easily recognized. In the first nymphal stage, bed bugs may be too small for some people to see well with the naked eye.
4.8 c	Male insects have a tendency to be smaller than females and bed bugs are no exception. Male bodies are smaller and narrower than the females.
4.9 a	Bed bug eggs are slightly curved cylinders that are attached to the substrate when deposited by the females. The nymphs escape the egg through an open cap at one of the ends of the egg.
4.10 c	Female bedbugs can deposit more than 500 eggs in their lifetime, which can make a population grow very fast, even when the infestation is initiated with a few individuals, or even a single female bedbug.
4.11 b	The bedbug lifecycle includes five nymphal stages, which differ from the adults by having a light colored cuticle that allows the red color of the blood to be seen more readily inside nymphal stages, especially the first two nymphal instars.

4.12 c	Typically, under good conditions and adequate access to blood meals, the bed bug can develop from nymph to adult in a little longer than a month.
4.13 a	Depending on the nymphal instar or adult stage, bed bugs can consume as much blood as six times its unfed body weight. Most of the volume ingested is expelled as feces soon after the blood meal as the bed bug eliminates the water in the ingested blood.
4.14 b	Typically, bed bugs will sit on the bed or other furniture where the human host is located, and will bite the host by extending its mouthparts into the host.
4.15 a	Because bed bugs sit on furniture while sucking blood from the host, it is common that several bedbugs will be biting the human host at about the same distance from the surface on which they stand. This causes several bites to be in a row on the body of the host
4.16 d	Bed bugs are not known to transmit diseases through their bite. In the past, research has demonstrated that bed bugs may carry different disease organisms on their body or ingested blood, but transmission of diseases has not been confirmed.
4.17 b	The bed bug legs lack soft pads that insects use while climbing smooth surfaces. Therefore, bed bugs can be trapped or prevented from reaching certain locations if there is a vertical barrier that is smooth enough to prevent the bed bugs from climbing it.
4.18 b	The two bed bug species *Cimex hemipterus*, the tropical bed bug, and *Cimex lectularius*, the common bed bug, have distributions that mostly do not overlap. Throughout the world, infestations of *Cimex lectularius*, the common bed bug, are more commonly encountered in human-occupied buildings.
4.19 d	Even without transmitting diseases, bed bugs can potentially cause several debilitating problems that make this insect a medically important infestation.
4.20 a	Bed bugs are capable of feeding on many different animal hosts, and have been observed feeding on several human pets, poultry and several other animals.
4.21 c	As is the case for other blood-sucking insects, heat, CO_2, and other odors associated with common hosts serve as cues for the bed bugs as they seek a blood meal.
4.22 b	The number of eggs laid by bed bug females depends on the nutritional status of the female bed bug, but females can lay several eggs per day.

4.23 b	The nutritional status of the female bed bug determines how many eggs the female can produce, so smaller blood meals result in smaller egg clutches.
4.24 c	Despite the name "traumatic insemination", the female bedbugs have adaptations to allow the insemination though its cuticle on the abdomen, and the male sperm is deposited in the spermalege, an internal female organ that receives and stores the sperm.
4.25 c	The typical bed bug smell is due to the alarm pheromone. In a bed bug infested location, this odor can alert humans and other animals of the presence of bed bugs.
4.26 a	Bedbugs are not particularly resistant to heat, so high temperatures are one of the methods used to control bed bugs.
4.27 d	In a location where heat treatment is applied, bed bugs will move into locations where they can escape the effect of high temperatures, and this may lead them into areas where there is poor penetration of the hot temperatures. In these locations, it is possible the bed bugs will survive heat treatments.
4.28 c	Many modern populations of bed bugs have been repeatedly exposed to several pyrethroid products and developed resistance to these products. The use of these products may be very inefficient in controlling certain bed bug populations.
4.29 a	Typically, in a bed bug infested location, all different stages of the bed bug life cycle are represented in the population.
4.30 b	As bed bugs digest blood, they produce feces that are dark brown to black in color, which serve as a telltale sign of the presence of bed bugs in a location, especially when these are located in and around sleeping quarters and furniture.

5   Flies

5.1 a	Flies in food handling establishments are usually split into small and large flies. The small flies live and breed for the most part inside food handling establishments. They are commonly coming from drains, sinks, and other wet areas. The large flies are strong fliers and are usually invading structures through doors, windows and loading docks. The most important control of flies is finding and eliminating their breeding areas. Knowing where they came from is a key element in fly control.

5.2 d	The two groups of flies are filth flies and biting flies. Filth flies are usually associated with garbage and feces. Biting flies take a blood meal. These flies are usually encountered outdoors and can be a tremendous nuisance, like mosquitoes.
5.3 c	House flies are important filth breeding flies. They are almost ¼" long and have four narrow black stripes on the thorax.
5.4 b	The house fly can complete its life cycle in 7–10 days. Garbage collection is usually weekly in urban areas so the fly life cycle is interrupted by removal of waste before house fly eggs can develop into adults.
5.5 d	Houseflies have sponging lapping mouthparts. They can only ingest liquids as adults.
5.6 c	Bottle fly larvae have been seen migrating from the walls of houses due to dead animals either in the walls or attic. Just before pupation bottle and blow flies will leave the carcass of dead animals in order to pupate in the soil or some other dry place.
5.7 a	Flesh flies retain their eggs in their bodies until the egg hatches. That way, when they find a suitable food source for larval development, they lay their larvae rather than eggs on the dead meat or flesh.
5.8 d	Flesh flies have three stripes on their thorax and a checkerboard pattern on the abdomen. Some also have a red tip to the abdomen. These flies can be problems around dog kennels, feeding on the decaying high protein dog food and feces.
5.9 b	Soldier flies are known to prefer liquid feces for development. They can be found in septic tanks and sewer lines. If there is a broken wax seal at the base of the toilet, soldier flies can migrate into the house.
5.10 d	The phorid fly or humpbacked fly is a small gnat that has a strong arch to its thorax giving it a humpbacked shape. These flies will run rapidly along surfaces instead of flying when disturbed. They are common in hospitals and nursing homes.
5.11 c	The phorid fly not only breeds in any decaying matter of high protein content, but it can also be associated with sewer and drain pipe leaks in slabs under food handling establishments.
5.12 a	When plants are overwatered, fungus can grow in the soil. Fungus gnats feed on fungus and are usually in the family Sciaridae. They are small and dark in color with long legs and smoky colored wings.

5.13 d	Drosophila flies are often called fruit flies or vinegar flies. The larvae feed mostly on yeasts found in fermenting fluids.
5.14 b	Small flies, like fruit flies, do not fly in moving air. A small fly problem generated by infestations in fruits and vegetables will not be noticed by customers. If they saw small flies in the produce section, they would not like the idea of buying infested produce.
5.15 a	Most filth flies have aristate antennae. The antenna actually has three segments: scape, pedicel and flagellomere (postpedicel). The arista is a hairlike structure that originates from the flagellomere.
5.16 a	Horse and deer flies have antennae with three segments. The scape, pedicel, and flagellum. On the tip of the flagellum there are a series of small segments called stylets.
5.17 c	Horse flies have a piercing lapping type of mouthpart. They first cut open the skin with their piercing parts. The blood flows out of the wound due to anticoagulants in the saliva. Then they lap up the blood.
5.18 a	Stable flies have piercing sucking types of mouthparts. The fly looks a lot like a housefly, but the proboscis is a thin rod that extends in front of the head.
5.19 d	Moth or drain flies develop in wet decaying organic matter usually found in drains, sewers and trickle filters of sewage treatment plants. The larvae are almost like mosquito larvae with spiracular openings at the end of a short siphon at the tip of their abdomen.
5.20 c	Even though fruit fly maggots often develop in rotting fruit and vegetables, the flies themselves are filter feeders and feed on yeasts that grow in the sugary solutions associated with decaying fruits and vegetables.
5.21 c	Fly light traps are effective in catching flies inside. They should not face the outside entrance doors because they can entice flies to enter the building.
5.22 c	Flies usually respond to light when they are within 20–25 ft of a light source. The lights should be placed as close to the floor as possible. Flies usually fly up to 5 ft above the floor. Fly catch is greatly reduced when lights are placed higher than about 8 ft.
5.23 a	Glue boards are the most frequently used device to catch flies in ILTs. Flies that are attracted to the light are often caught in the glue board that is placed behind the light. Those boards are usually either white or black in color.

5.24 b	ULV stands for ultralow volume application of insecticides. An ultralow volume application is a space spray used to knock flying insects, like flies and mosquitoes, out of the air.
5.25 d	House flies were one of the first insects to develop insecticide (DDT) resistance. They have continued to develop resistance to even the newest insecticides.
5.26 c	Standard window screen is 18 by 16 mesh and is typically used on window and doors of houses. Screened enclosures for porches and pools usually have an 18 by 14 mesh size. The term mesh stands for the number of openings per inch.
5.27 b	Z-9-tricosene is the fly sex attractant used in many fly baits. Some of the spot baits and scatter baits utilize the product to attract male flies for sex.
5.28 a	Fly baits are usually used outdoors. The most frequent kind of fly bait is a scatter fly bait that is in the form of granules. The granules are scattered on areas where flies are frequently found.
5.29 d	The fungus gnat is a small fly that can live in overwatered plants in offices and homes. If the soil is overwatered and not allowed to dry between waterings, fungus gnats can become a problem.
5.30 b	Emptying dumpsters, cleaning up trash and garbage, and removing manure from the area are all methods of house fly control. These measures will not affect stable flies. The main sources of stable flies are hay and silage that is left on the ground and mixed with animal manure.

6 Cockroaches

6.1 b	Tarsal pads and arolium are fleshy adhesive pads on the tarsus of cockroaches. These provide traction on smooth surfaces. They also are thin parts of the exoskeleton that are permeable to residual insecticides. Cockroaches crawling on surfaces pick up insecticides through the tarsal pads and arolium.
6.2 a	Cockroaches lay their eggs in egg capsules called oothecae. There may be an average of 10–30 eggs laid in each ootheca.
6.3 d	Cerci have fine hairs on their surface for air movement and vibration. Cockroaches can detect a person entering a room before a light is turned on by detecting the air movement of a door opening or a person moving. Cerci in contact with a surface can detect vibrations.

6.4 d	Fat bodies of well fed cockroaches take up most of the interior body of the cockroach. They are composed of three types of cells: trophocytes that are the main storage for lipids (oils) which are a source of energy, urate cells that store uric acid which is the main nitrogen excrement for terrestrial insects, and mycetocytes which convert the stored uric acid into usable amino acids for protein synthesis.
6.5 b	The foregut and the hindgut of cockroaches arise from the embryonic ectoderm resulting in the surface being covered with chitin. The midgut arises from endoderm which does not have a covering of chitin. Therefore, the only part of the cockroach gut that allows food to pass through for nutrition is the midgut.
6.6 a	Malpighian tubules are the main excretory organ of the cockroach. They are diverticula (evaginations) of the midgut and empty collected metabolic waste from the insect's body at the pyloric valve.
6.7 d	Cockroaches are considered primitive termites, linked with the wood-feeding cockroaches (Cryptocercids) and the most primitive termites (Mastortermidae). All cockroaches and termites have proctodeal trophallaxis which is the passing of resources to other members of their species group through their feces. Adult cockroaches defecate for first stage nymphs to ingest.
6.8 b	Urban cockroaches defecate in cracks and crevices (harborages) close to food and water. These harborages are marked with cockroach feces which contain the aggregation pheromone. Cockroaches can defecate while running or walking, which means that there can be fecal trails for them to follow to food and water.
6.9 c	Cockroaches are known to mechanically transmit disease causing organisms. In other words, they pick up pathogens from the environment and transfer them to the food or skin of humans. Many of these are bacterial, associated with food poisoning and skin infections.
6.10 b	The German cockroach has been extensively studied and linked to allergic responses in sensitized children and adults. The allergens that have been identified are Bla g 1 to 12. Bla g 1 is the most studied and is associated with digestion in the cockroach midgut.
6.11 a	A standard vacuum cleaner is NOT recommended for removal of cockroaches and their debris from infested locations. The filters in the bags will not remove cockroach particulates which can become airborne during cleaning operations. A vacuum with a high efficiency particulate air filter (HEPA) is recommended for any work with cockroach or cockroach debris removal.

6.12 c	Cockroaches hide in cracks and crevices. The ear canal of a sleeping person is a great location for cockroaches to hide. After they enter the ear canal, cockroaches cause tremendous pain inside the ear. Although cockroaches can gnaw on skin and eyelashes, the incidence of this occurring is minor compared with cockroach invasion of ear canals. Cockroaches have secretions that make them smell and taste bad.
6.13 d	Courtship in German cockroaches is well-studied. It begins with females calling males by releasing a long-range sex attractant, blattaquinone. They then antennate which is a process of assessing each other. The male then raises his wings and emits a sex pheromone from his tergal gland. The mounting process is initiated by the female feeding on the male's tergal gland secretion.
6.14 c	When cockroaches are crowded, the nymphs and adults produce salivary compounds that mark shelters as unsuitable. This results in males and nymphs dispersing most readily to other locations where they are not overcrowded.
6.15 d	German cockroaches do not fly and their infestations are associated with human structures. Genetic studies of German cockroach populations show that infestations in structures are spread by long- and short-range human movement of cockroaches.
6.16 b	Surveys of people in various parts of the world have documented that cockroaches are uniformly disliked. In fact, we are evolutionarily programmed to dislike them.
6.17 a	Cockroaches survive in a wide variety of conditions. They are found in the Arctic and in deserts. Cockroaches have been reported to survive at $-122°C$ by producing glycol (antifreeze) in their bodies.
6.18 d	Hissing cockroaches produce sound with the use of spiracles on the fourth abdominal segment. The spiracles are able to close while the cockroach contracts its body muscles to build up hemolymph (blood) pressure.
6.19 c	American cockroaches now occur worldwide. They were transported on ships and were widely associated with filth and lack of sanitation on early sailing vessels.
6.20 a	Cockroaches do not digest the materials used to make clothing. They do not digest cotton or wool. In the process of feeding on starches and oils on clothing, they can cut holes in fabric.

6.21 c	Cockroaches are known to be scavengers feeding on a wide variety of plant and animal matter.
6.22 d	Male cockroaches have a pair of styli at the posterior tip of their abdomen. Females lack styli. Only female cockroaches will carry or develop an ootheca.
6.23 b	About 75–80% of natural cockroach populations are nymphs. Small nymphs are usually the most prevalent stage caught in sticky traps placed in the field for monitoring.
6.24 b	Food foraging for cockroaches is often observed just after sunset. However, females carrying egg capsules do not move much and do not feed often. Also, first and second stage nymphs do not move often and may forage at virtually any time of day or night.
6.25 c	Female cockroaches exhibit parental care for young nymphs. They will produce feces to transfer nutrients to nymphs as well as beneficial microorganisms to assist in food digestion. Females will not finish copulating by eating male cockroaches.
6.26 a	After nymphal cockroaches molt to the next stage, it takes a few hours for the new exoskeleton to harden and darken.
6.27 c	
6.28 b	Long filamentous antennae are characteristic of cockroaches. Although antennae are used for all the functions listed, it is thought that the use as a tactile organ is most important.
6.29 d	Cockroaches have two large compound eyes and three simple eyes (ocelli) on the head. The compound eyes are located anteriorly and laterally, wrapping around the sides of the head. The cockroach head is tucked under the pronotum and the mouth faces posteriorly.
6.30 c	The three large nerve ganglia in the cockroach are located so that one is in each of the three thoracic segments. A cockroach can start running before the nerves in the head are aware of stimulus. The thoracic ganglia are responsible for movements of the legs and wings.

7   Subterranean Termites

7.1 a-1 and b-2	These features clearly distinguish termite alates from ant alates. In addition, antennae of termite caste members are always straight, whereas in ants it is mostly bent antennae.
7.2 a	Unlike ants which are often found in similar niches as termites, the three body parts, namely head, thorax and abdomen of termites, are broadly joined without any constrictions. In ants these three parts are clearly distinguishable.

7.3 a and b	Termites depend upon the microbes in their gut to digest the complex sugars in wood into simpler molecules. In this process the microbes also produce gases such as methane.
7.4 b	Workers and soldiers live side by side in a colony. Soldiers are distinguishable from workers by a much-enlarged head. The head has large mandibles, except for nasutiform soldiers which lack mandibles and have a head prolonged in the form of a nasut.
7.5 c	Isoptera originates from two Greek words, "iso" meaning equal and "ptera" meaning wings. This refers to the similar size, shape and venation of the four wings present in termites.
7.6 b	Termite queens demonstrate extreme physogastry or the expansion of the abdomen to an extreme size. This phenomenal growth is largely due to distended ovaries and fat deposits inside her abdomen, which aid rapid egg laying.
7.7 a	Even though the species are known as subterranean, they can still construct nests away from soil. These are generally constructed in tree holes, stumps and even structures. The critical part of these above ground nests is the presence of a constant moisture source.
7.8 e	The reason termites are called social is because they live in an organized society. They have one or a few females responsible for all the egg laying, while other members of the colony (usually sterile females) gather food and do other tasks.
7.9 b	This genus is unique due to the soldier's nasut (an elongated frontal projection on the soldier's head). Soldiers use the nasut as a defense against ants and even large vertebrates such as ant eaters, by squirting a noxious defensive secretion.
7.10 a-3, b-4, c-2, d-1	
7.11 b	One characteristic of Formosan subterranean termites, *Coptotermes formosanus* as well as other species belonging to genus *Coptotermes*, is producing carton nest material that is made of termite excrement, chewed wood and soil.
7.12 b	During cold months, social insects like subterranean termites go into a quiescent state, huddling together until warmer weather returns.
7.13 a and d	The lower termites predominately eat wood. On the other hand, higher termites consume a wide variety of materials, including animal feces, humus, grass, leaves, and roots. The gut in the lower termites contains many species of bacteria along with Protista, while the higher termites only have a few species of bacteria with no protistans.

7.14 a	There are 28 species belonging to three families, namely Rhinotermitidae, Termitidae and Kalotermitidae which are invasive. The characteristics of invasive species are: they eat wood, nest in food, and easily generate secondary reproductives. These are most common in Rhinotermitidae and Kalotermitidae among the three families.
7.15 d	*Coptotermes formosanus* was first described from Taiwan in the early 1900s, but is a native to Southern China. Today the species is found in Mauritius, Réunion Island, and across the Pacific Ocean to parts of Polynesia, Hawaii, Marquesas Island, Micronesia, Fiji, Mexico, Florida, the Caribbean Islands, and is now spreading along the south Atlantic coast of Brazil
7.16 a and b	Moisture may enter structural wood as a result of condensation, contact with soil, or leaks. Termites prefer to attack wood which is moist. Thus, controlling moisture is a useful method for preventing termite attack.
7.17 a	Termites are prone to desiccation. A continuous supply of moisture either from wood or moist surroundings is a must for survival.
7.18 a	Trophallaxis is the exchange of food between two individuals. The food exchanged may be salivary secretions or regurgitated gut contents. In termites, proctodeal trophallaxis is crucial for replacing the gut endosymbionts that are lost after every molt.
7.19 e	Termites construct mud tubes for shelter as they travel between their subterranean nest and the above ground feeding area located in a structure. Due to poor abdominal sclerotization, they are susceptible to desiccation and need a mud cover to remain protected.
7.20 d	A mature colony may cause severe structural damage in a very short time. For this reason, it is extremely important that an infestation when detected should be quickly taken care of.
7.21 a-4, b-2, c-1, d-3	
7.22 b	One of the methods for keeping termites from infesting a structure is to create a chemical barrier. The chemicals used have a long residual activity in addition to being repellent in nature. However, it is important to choose the right formulation with the correct active ingredients.
7.23 a and b	Bifenthrin is a synthetic pyrethroid insecticide. It is a repellent that prevents foraging of termites in the treated area and thus maintains a barrier against attack by termites around structures.

7.24 a-2, b-1, c-2, d-2	Repellent and non-repellent are relative terms and have been popularly used in the industry to categorize termiticides. Some researchers consider the terms arbitrary, but their use remains popular due to observable distinctions in behavioural responses to the chemicals by termites.
7.25 b	The poisoned termites in the treated zone unknowingly share the termiticide with other nest mates through social interactions; this can cause secondary mortality in nest mates that have not been directly exposed to the treated areas.
7.26 a	Horizontal transfer of insecticide occurs when insects contact or ingest an insecticide, return to a nest, and transfer the insecticide to other conspecific insects through contact. This phenomenon can facilitate the spread of insecticide into cryptic populations such as termites.
7.27 a, c and d	Termite baiting is a method of eliminating termite colonies with the use of a food-based bait. The bait stations are installed in such a manner that it intercepts active termites around the structure. Once termites are aggregated in the station, they are fed on a bait containing an insect growth regulator, leading to colony elimination.
7.28 e	Subterranean termites entering through concealed entry points usually do not leave any signs of their presence. Thus, inspecting all wood in the structure, particularly by tapping, gives a good indication if it is being infested and hollowed out from inside.
7.29 a, b and c	A building may be made of concrete and bricks, but it will not prevent termites entering the structure in search of wood, which may be in the form of furniture, wall cabinets, etc. Termites are also known to look for other cellulosic materials such as paper, fabrics, etc.
7.30 a and d	Termites are known to damage many types of materials; thus, choice of building material is important in construction, particularly in termite prone areas.

8   Drywood Termites

8.1 b	Drywood termites can exploit wood as low as 3% in moisture content. They have special mechanisms to conserve water in their body.
8.2 c	The colonies are usually small, with numbers ranging between a few hundred to thousands. With the passing of time the colony faces serious depletion of food which limits colony size.

8.3 b	Drywood termites have no contact with soil. They depend on the humidity-rich air of coastal areas in setting up their colony.
8.4 b	A total of 456 species are associated with the family Kalotermitidae, which represents all drywood termites. But the family includes species with a diversity of habitats such as dry and damp wood, as well as subterranean species.
8.5 a	There are 21 genera currently recognized within the Kalotermitidae; however, studies on the biology and ecology have been restricted to five including: *Cryptotermes, Glyptotermes, Incisitermes, Kalotermes* and *Neotermes*.
8.6 a, b and d	25% of the 28 listed invasive termite species belong to the family Kalotermitidae. Their ability to nest in the wood they eat among others makes them truly invasive.
8.7 a	Colony fusion is known to occur after two colonies in the same resource encounter each other.
8.8 b	The presence of pelletized, hard, ~1 mm long fecal matter distinguishes drywood termites from other wood infesting insects.
8.9 c	The subterranean termite has a single vein running parallel to the top edge of the wings. The drywood termite has numerous, more complex veins running between the parallel vein and edge of the wing.
8.10 b	Drywood termites do not require ground contact and do not build mud tubes.
8.11 a	
8.12 d	There are limitations to each technique/device. All have been tested to varying degrees, and have been reported to have some usefulness for easy-to-reach infestations that are exposed.
8.13 a-2 and 4; b-1 and 3	
8.14 c	Drywood termites have the ability to metabolize water from the wood that they eat, absorbing and reabsorbing water from their feces as needed.
8.15 b and c	Subsurface injections employing the drill-and-treat method and fumigation are the popular commercial methods used for drywood termite control.
8.16 a	The mode-of-action in killing drywood termites using excess heat is complex and involves hyperthermia at the cellular level leading to the disruption of cell membranes and destabilization of enzymes.

8.17 c	Eight species of drywood termites are known through the published literature to be eliminated by use of sulfuryl fluoride (SF).
8.18 a	Methyl bromide has been implicated in depletion of the ozone layer and is therefore now restricted to use in quarantine-type situations, and a few agricultural commodities.
8.19 a and b	When applying fumigants, it is essential to know the level of concentration above which it is not safe to subject workers and also the maximum periods of exposure, including repeated exposures during normal working hours. Such concentrations are widely known as threshold limits and are usually expressed in terms of parts per million by volume in air.
8.20 e	Severe poisoning may result in fluid in the lungs. This can lead to dizziness, blue or purple skin color, unconsciousness, and even death.

9   Powderpost Beetles and Wood Pests

9.1 d	Both the presence of small holes that are round or oval and the presence of sawdust in the beetle galleries are necessary signs for determination of the presence of beetles in contrast with other insects that may attack wood.
9.2 b	Each beetle family has its preference for the type and sometimes age of the wood they attack. Therefore, clear determination of the type of wood under insect attack is an important step in the identification of the wood destroying organism.
9.3 b	For those beetles that attack dying trees, once their presence is determined in processed wood, they will not be able to initiate a new attack unless they are given access to living trees again.
9.4 a	Once insects have emerged and left the galleries in the wood, they will likely not be seen again, so the observer is left with the characteristics of the gallery in order to attempt identification of the attacking insect.
9.5 b	Because the development of wood-infesting beetles is very slow, and due to the nutrient restriction in the diet, wood is still considered new after 7 years.
9.6 c	For the purposes of investigation of wood-infesting beetles, a wood is only considered old after 10 years have passed since it was processed in a mill.

9.7 c	Hardwoods have large pores and no resin ducts; they often have more distinctive grain patterns than soft woods and often contain light-colored sapwood and dark heartwood. Softwoods have horizontal and vertical resin ducts and no pores.
9.8 c	The wood from these trees has a very distinct grain pattern, and once dry, has a tendency to be heavier and harder to penetrate with nails.
9.9 b	The wood from these trees does not have a very distinct grain pattern, and has large pores which tends to make it softer and easier to penetrate with nails.
9.10 c	Hardwoods are often used in decorative wood pieces where the accented difference between hard and softwood makes for better decorative effect.
9.11 a	Softwoods are often used in wall framing because they are light, relatively inexpensive, and easy to use in construction.
9.12 d	The term "powderpost beetle" applies to any of the wood-boring species of three closely related groups.
9.13 c	Recently, these groups have been reclassified. The Bostrichidae have the traditional family Lyctidae now as a subfamily, Lyctinae, along with the subfamily Bostrichinae. The Anobiidae is a subfamily of Ptinidae (Anobiinae).
9.14 b	Anobiids are unique among beetles frequently attacking seasoned wood because individual species feed on both softwood and hardwood.
9.15 c	Adult bostrichids often have rough texturing on the front of the pronotum and sometimes on the rear margins of the elytra.
9.16 a	Lyctid powderpost beetles usually infest furniture, flooring, paneling and molding.
9.17 d	For lyctid control, only hardwood products need to be treated. For small infestations, removal and replacement of infested products is recommended.
9.18 c	Bostrichids are commonly found in bamboo, wicker and other natural basket material.
9.19 b	Cerambycids larvae are commonly called round headed borers and can be quite long (>50 mm) and make large galleries in the wood as they develop. Because the development takes so long, sometimes they may still be feeding on the wood after it becomes part of a building.
9.20 a	The longhorned beetle known by the common name "old house borer", *Hylotrupes bajulus*, may attack timbers in a building thus requiring control measures.

9.21 d	Cerambicidae is a very large family of beetles mostly characterized by their extremely long antennae, which in many cases extend well beyond the length of the beetle body.
9.22 a	Wood-attacking insects need relatively high temperatures for optimum development, and can be controlled by placing infested materials under freezing conditions for several days.
9.23 b	Wood-infesting insects get into homes, most commonly when people move infested furniture or other objects into the buildings. Making sure that wood objects are not infested, or treating any wood material in a way to remove any infesting pest, are the best ways to prevent wood pests from getting established in buildings.
9.24 a	Making sure that wood in a building or any furniture maintains a moisture level below 10% will prevent the development of wood infesting beetles.
9.25 c	Borate products have been used in the treatment of wood to prevent insect damage. It is approved for used both as a pre- and post-construction product that will kill insects that feed on the treated wood.
9.26 d	Borates can leach into water, so wood in contact with the ground and liquid water eventually will lose the borate content necessary to prevent insect development.
9.27 b	Sealing products fill in the wood pores and beetles are not able to find adequate oviposition spots, preventing beetle populations' growth and establishment.
9.28 a	Concrete floors and plastic barriers prevent moisture from the soil from getting trapped inside the basement and consequently into the wood in the house structure. Keeping wood very dry ensures that the wood in the structure stays below the moisture levels required for insect development.
9.29 d	Heat treatment must ensure not only that a killing temperature is reached, but also that the treatment is long enough for that temperature to penetrate the whole structure and any timber present in the location.
9.30 a	The typical staining seen when ambrosia beetles attack wood will decrease the value of the wood and in many cases make some portions of the wood unusable.

10 Ants

10.1 b	In most ant species, each colony is restricted to a single nest location, but some species are known to have a single colony occupying numerous nest sites, and each of these sites may have several queens in the polygyne species.
10.2 c	While in most ant species a single queen is the norm, many species, especially the most common pest ants, may have more than one reproductive queen, and in some polygyne nests hundreds of queens may be present.
10.3 c	After mating, normally the males will die or be killed and the female reproductive will find a suitable location to lay eggs and rear the first brood to start a new colony.
10.4 d	Budding refers to the formation of a new colony by a simple method of splitting the original colony into new colonies that will then operate as a new entity completely separate from the original colony.
10.5 a	The best method to avoid ant problems is to prevent their establishment in a location, mostly by preventing the attraction of ants to a location by elimination of food and water sources.
10.6 b	Aphids and other sap-sucking organisms are a major source of sugar (energy) for many ant colonies; therefore, the elimination of these organisms helps in the elimination of ant problems.
10.7 a	Sticky substances that trap ants or prevent their passage are a major disrupter of foraging behavior that will prevent the ants obtaining food.
10.8 b	Baits are taken into the nest and spread among the entire population of ants, killing adults and juvenile stages.
10.9 b	Cracks and crevices in walls are one of the main ways that ants can get into buildings, so closing these openings is an important part of prevention.
10.10 c	Plants attract ants close to buildings but also provide a bridge over any pesticide application that may have been placed around the outside perimeter of the building.
10.11 d	Because many sucking insects are an important source of sugar (in the form of sugary secretions they produce) for ants, elimination of these insects cuts down on the nutrients ants can take to the nest.
10.12 b	Potted plants can serve as a hiding or nesting place for ants, therefore their elimination prevents nesting by ants very close to or in buildings.

10.13 b	Elimination of foraging ants only targets a very small percent of the ants in the nest, while application of baits that reach the queens and the larvae will have a stronger effect by preventing further colony growth.
10.14 a	Baits that ants recognize as food sources are taken into the nest and fed to the larvae, causing mortality in immature ants before they can go out and forage.
10.15 d	A filtering mechanism in the digestive tract of adult ants prevents the ingestion of solid particles, therefore solid particles are fed to larvae.
10.16 c	Adult worker ants do not have wings, and only reproductive individuals have wings which allow male and female adult reproductives to fly out of nests to mate.
10.17 b	Trophollaxis is the process by which food, and therefore toxic bait material, is shared among the members of an ant colony.
10.18 c	The ideal bait should be non-repellent so the ants are not driven away from it, and on a preferred food so the ants will collect and consume it. A slow active ingredient guarantees that the material gets distributed and consumed widely in the ant nest.
10.19 d	Bait stations protect the bait from the elements, and allow the ants to consume it for a long time.
10.20 b	Rain can destroy ant bait granules and wash the bait away from its intended target.
10.21 a	Timing of ant application should coincide with the time the ants are starting to look for food, so there is a quick discovery and maximum consumption of the bait.
10.22 c	The sweet residue in soft drink containers are a source of attraction for ants.
10.23 d	Sticky surfaces prevent ant movement and foraging for food.
10.24 b	Soapy water is a cheap and safe way to kill ants without concerns for the safety of children and pets.
10.25 c	Early in the spring, there are fewer food sources for the ants so sweet baits are very attractive and efficient.
10.26 b	Slow-acting ingredients work best for ant control because most of the ants do not forage outside the nest, and the active ingredient has to be widely distributed in the nest to be effective in eliminating the ants.
10.27 d	Blocking re-entry of ants limits the population indoors, so that the control actions will be more efficient in eliminating ants that cannot be replaced by new ants coming in from outdoors.

10.28 b	Sugar-based bait is not very effective with thief ants because they prefer greasy foods, meats, cheeses and similar animal products.
10.29 c	Because pharaoh ant colonies have hundreds of queens, each fragmented colony continues to grow and soon the location is completely taken by the ants again.
10.30 d	Despite quarantine efforts, fire ants (*Solenopsis invicta* and *Solenopsis richteri*), have been introduced in several areas of the world and cause problems where they exist.

11   Fleas, Ticks and Mites

11.1 a	The cat flea is the main species of flea found on cats and dogs in the US. Some people get confused because the cat flea can be found on many species of animals, including dogs.
11.2 d	The oriental rat flea is the main species of flea that transmits plague. It is usually a parasite of rats, but when rats are controlled, the flea can transfer to humans. If the previous rat host was infected with plague, then the oriental rat flea can transmit the disease to humans.
11.3 c	The human flea *Pulex irriitans* and its closely related species, *Pulex simulans*, are found throughout the world. The pet trade has probably distributed these species to most parts of the world. They are prevalent species on animals associated with urbanization, like opossums and raccoons. The will bite humans and pets like cats and dogs.
11.4 c	Fleas that feed for a long time at a site seem to be burrowing into the skin, but they are really just surrounded by irritated skin that swells and surrounds the flea. This medical condition is called tungiasis, and is different than the allergic reaction caused by flea allergy.
11.5 b	Fleas feed by injecting saliva into the skin. The saliva can cause irritation and an allergic response in cats and dogs. The allergic response is called flea allergy dermatitis.
11.6 d	When fleas ingest the bacterium for plague, *Yersinia pestis,* the flea gut becomes blocked because of the bacterial growth. Due to the blockage, the flea tries to get additional blood meals, but regurgitates the bacteria into the host.
11.7 b	Tapeworms in the intestine of cats and dogs produce proglottids, egg sacs, that are filled with eggs. These proglottids crawl out of the anus of the infected host. They are eaten by flea larvae which are not affected. When the larva molts to the adult, the host can ingest the infected flea, and the tapeworm then can infect the host.

11.8 d	Cat fleas lay eggs on the host. Those eggs are not attached and readily fall off the host when the host moves or scratches. The rat fleas remain in the rodent nest and wait for a host to feed on.
11.9 b	Flea larvae feed mainly on the adult flea feces. Adult flea feces is mainly dried blood. It dries on the host and falls off into the same environment as flea eggs.
11.10 c	Fleas remain as pre-emerged adults waiting for a host to come close to their cocoon. They can remain alive in the cocoon for 6–12 months. The carbon dioxide, vibrations, and heat of the host stimulate the adults to emerge from the cocoon.
11.11 d	The flea cocoon does not allow sprays to contact the pre-emerged adult inside. Cocoons have been dunked in spray solutions, and adult fleas still emerge. The best control is usually to stimulate adult emergence and then spray the adult fleas with insecticide.
11.12 d	Fleas are important vectors of disease organisms, like plague, murine typhus and cat scratch fever. Mosquitoes are the vectors of dog and cat heartworm (Dirofilaria immitis). Fleas are not considered vectors of heartworm. Dirofilaria immitis has been found to occur in fleas, but it has never been proven that fleas can vector the pathogen.
11.13 a	Ticks use their Haller organ located on the first pair of legs to detect hosts passing nearby. Ticks crawl up vegetation and quest for hosts by waving their front legs while holding on with their other legs. When a satisfactory host brushes past them, they grab hold and attach to the host.
11.14 b	Eggs of ticks hatch, and a larva emerges. The larvae of ticks have six legs, and all the other developmental stages have eight legs. Also, ticks do not have wings, whereas most insects are winged for greater movement and dispersal in the environment. Ticks mainly rely on their hosts to be dispersed.
11.15 d	The order Acarina (class Arachnida) includes mites and ticks. Members of this order differ from other arachnids in that the body is not segmented, and the cephalothorax and abdomen are combined into one body region.
11.16 d	Ticks are great vectors of pathogens because they have a tendency to feed on several hosts during their development to adults. Most ticks will feed for several days on the host. This allows the passage of pathogens between the host and the tick.
11.17 b	Hard ticks have a hard upper surface called a shield or scutum that covers the entire back of the male but only partly covers the female. They are in the Ixodidae family and are the main ticks for transmission of diseases.

11.18 a	The American dog tick is a tick that can paralyze its host with its saliva. A tick that embeds along the spine or on the skull can paralyze its host to prevent it being removed by grooming.
11.19 c	The black-legged tick, *Ixodes scapularis*, is the main vector of Lyme disease in the eastern US. That is where most of the cases occur. On the west coast of the US, the main vector is the western black-legged tick, *Ixodes pacificus*.
11.20 c	The Lyme disease bacterium, *Borrelia burgdorferi*, is spread through the bite of infected ticks. Most humans are infected through the bites of immature ticks called nymphs.
11.21 c	Blacklegged ticks are considered three-host ticks. Adult females drop off the third host to lay eggs after feeding, usually in the fall. Eggs hatch into six-legged larvae that seek out and attach to the first host, usually a small rodent. Engorged larvae leave the first host and molt into nymphs. The nymphs seek out and attach to the second host usually another rodent or bird. The nymphs feed on the second host, usually a deer, dog, or human. The adults seek out and attach to a third host, which is usually a larger herbivore (deer, dogs, and humans). The life cycle usually takes 2–3 years.
11.22 b	Brown dog ticks are quite capable of completing their life cycle entirely indoors. They are one of the only ticks that actually hunts for a host, rather than quest for it. It has the ability to sense the host and then crawl towards it so it can attach and feed.
11.23 a	An engorged female tick can lay more than 5000 eggs in cracks and crevices and under furniture. When they hatch, thousands of tick larvae, called seed ticks, will crawl up onto furniture and up the walls. They can wait for months without feeding, waiting for a host to come close. Then the seed ticks will crawl onto the host and start feeding on the host's blood.
11.24 a	The house dust mite occurs in human dwelling and feeds on shed skin. It produces allergens that become airborne and cause respiratory problems and asthma.
11.25 c	Human scabies is caused by a skin infestation by the human itch mite (*Sarcoptes scabiei*). The female scabies mite burrows into the upper layer of the skin where it lives and lays its eggs.
11.26 b	Straw itch mites feed on insects, often killing them. They can be a problem when dried foodstuffs become infested with the larvae of storage insects in warm, humid environments.
11.27 d	The northern fowl mite is a widely distributed mite on wild birds and poultry. Northern fowl mites usually infest houses from wild bird nests. When the young birds leave the nest, the mites migrate into the house and attack humans.

11.28 a	Ticks and mites differ in many ways. Ticks are larger and tend to be 3 mm or more in length. Mites are smaller.
11.29 c	Ticks and mites are in the class Arachnida which includes spiders and scorpions, mites, and ticks. Those arthropods have a segmented body divided into two regions of which the anterior bears four pairs of legs but no antennae.
11.30 b	The hypostome bears the mouth opening on mites and ticks. Mites do not have barbs on their hypostome, whereas ticks have barbs that prevent the easy removal of ticks from the host.

12   Sporadic Pests

12.1 b	The most common lady beetle to enter houses in the US is the multicolored Asian lady beetle. The beetle is native to Asia and China and now occurs in large numbers throughout the eastern US. Asian lady beetles outdoors feed on aphids and scale insects.
12.2 c	Plaster beetles get their name from their tendency to appear after applications of plaster to the walls and ceilings of houses. The curing plaster releases large amounts of moisture that stimulates the growth of molds and mildews (fungus). The beetles thrive in these high moisture conditions and feed on the fungi that develop indoors.
12.3 a	Springtails are the most frequently collected insect around the perimeter of structures. They eat molds and mildews and are present in moist areas. In fact, thousands have been caught in sticky traps placed on the ground. Springtails have a structure on their abdomen that springs them into the air when they are disturbed.
12.4 d	Thrips have hairs on their wings that allow them to fly. Most insects have a membranous wing with veins. Thrips usually feed on flowers and suck fluids from plants for nutrition. During dry weather, thrips are attracted to the sweat on people. They have a rasping mouthpart, and even though they are really small, their attempts to feed on sweat are really painful.
12.5 a	Mayflies are in the order Ephemeroptera and usually have one generation per year. Because they only have one generation per year, they emerge in large numbers, sometimes covering the sides of buildings, roads, and the ground.
12.6 d	Aquatic midges look like mosquitoes but do no feed in the adult stage. So, they do not bite humans and animals for a blood meal. They only live as adults for a day or two. Around the world, aquatic midges emerge in huge numbers covering the sides of buildings and vegetation near water.

12.7 c	Moth caterpillars develop by feeding on plants. Fall webworms, Eastern tent caterpillars, gypsy moths and oleander caterpillars will feed and develop on their host plants. But when it is time for them to pupate, they leave the plant and migrate in large numbers looking for a place to spin their cocoon.
12.8 c	Urticating hairs on the body of caterpillars protect them from their natural enemies. The hairs are hollow and really sharp. When people touch them, the hairs break off in the skin and release a poison that is painful and irritating.
12.9 a	Millipedes are in the class Diplopoda, meaning that they have two pairs of legs per body segment. Their body is usually cylindrical and long, with many segments. Millipedes feed on decaying organic matter such as decaying leaves and vegetation.
12.10 d	Centipedes are predators and feed on a lot of the small animals in their environment. They differ from sowbugs, pillbugs and millipedes that feed on decaying vegetation. Centipedes have many segments with one pair of legs per body segment.
12.11 b	Sowbugs and pillbugs are crustaceans that have developed the ability to live on land. Many other crustaceans like shrimp and crabs spend almost their entire lives in water. Sowbugs and pillbugs have oval bodies with hard overlapping plates covering their head and body segments. They feed on decaying vegetation and organic debris.
12.12 d	Lawn shrimp are in the class Amphipoda. The ones that migrate into houses are terrestrial and are usually found in litter under trees and shrubs. During heavy rains, the lawn shrimp will jump or crawl into garages and houses. They are brown in color when they are alive, but after death, they turn into a typical pink shrimp coloration.
12.13 c	Brown marmorated stink bugs use their proboscis to feed on plant tissue. They are an introduced species in the US and have been invading houses in huge numbers. They emit a strong odor by squirting chemicals from their abdomen in defense of being eaten by birds and lizards. These insects should not be vacuumed because that disturbance causes the release of a lot of noxious odors.
12.14 a	Boxelder bugs are a black bug with the edges of the wings and thorax rimmed with a bright red color. Boxelder bugs feed on the seeds of maple, boxelder and ash trees. They can be present in large numbers that invade structures in the fall to search for an overwintering location.

12.15 b	Almost all species of scorpions found in and around houses in the US are not considered highly dangerous. Most have a sting that is no more problematic than a wasp sting. Most scorpions live under the bark of logs and dead trees. When people pick up firewood to build a fire to cook or to stay warm, scorpions can exit from under the bark and sting the person handling the wood.
12.16 a	Scorpions have four pairs of legs. The one pair of pedipalps is enlarged as pincers for holding onto prey. The last segment of the abdomen is bulbous with a stinger that inflicts a painful sting. Scorpions are predators and usually use the sting and pincers to immobilize and hold their prey.
12.17 c	The whip scorpion secretes acetic acid which smells like vinegar. Therefore, whip scorpions are commonly called vinegaroons.
12.18 b	Tarantulas are a group of large hairy spiders that belong to the family Theraphosidae. All tarantulas are venomous. Their venom is not deadly to humans, but some bites are uncomfortable and might persist for several days. Tarantulas have urticating hairs and flick them off when they are upset.
12.19 d	Bark beetles may develop under the bark of firewood. These beetles will not infest the structural wood of houses and are only a nuisance. The bark beetles feed in galleries on fungus that they cultivate in living trees. When the tree dies, these beetles cannot survive long. If the firewood infested with these beetles is brought into the house, the beetles will leave the wood and disperse.
12.20 c	A velvet ant or mutillid wasp is usually brightly colored red and black as a warning coloration. They have a long stinger that inflicts a very painful sting when the insect is handled. In general, velvet ants do not require control and are not commonly encountered by customers.
12.21 c	Dermestid beetles, like hide and carpet beetles, are infrequently found pests in most houses. Dermestids feed on animal products like hides, stuffed animal heads, woolens, and other items made from actual animal materials. The larvae are covered with hairs and are characteristic in appearance. Their shed skins and hairs have been known to cause allergic reactions in people.
12.22 d	Clothes moths consume animal fibers, especially woollen items such as sweaters, carpets, rugs, and accumulations of pet fur. There have been several cases where severe infestations of clothes moths were associated with long-haired cat fur accumulating under and behind furniture.

12.23 b	Buprestids are metallic wood boring beetles that are shiny and brightly colored. The larvae are associated with the living sapwood of trees and may be a pest of log homes and rustic furniture and fences. Larvae prefer wet wood and may live in seasoned wood but do not reinfest structures.
12.24 c	The periodical cicada has either a 17-year or a 13-year life cycle. In 2021 the 17-year cicada emerged and made a lot of noise. As many as 20,000 to 40,000 cicadas have emerged from a single tree.
12.25 a	The Triatomine bugs, like the blood-sucking conenose bug, are important disease vectors in South America. They can transmit Chagas disease which is a slow developing disease that affects the smooth muscles such as the heart. Many people affected by Chagas die of a heart attack about 30 years after contracting the disease.
12.26 d	Leafhoppers are the largest family in the true bug order Hemiptera. They are associated with a wide variety of plants and have many colors depending on the species. Many species of leafhoppers are attracted to lights. Outside lighting brings them to the structure, and then they find small places to inter, perhaps being attracted by interior lights, as well.
12.27 a	Cicada killer wasps are not social insects, but colonial insects. They are large wasps sometimes over 1½ inches long. The wasps catch their prey, like cicadas, and sting them in order to paralyze them. These wasps do not usually attack people, but residents are often afraid of them and want them controlled.
12.28 d	Carpenter bees excavate holes in wood to build their nest. They are solitary bees and are not aggressive and rarely sting. The openings to the nest are about ½ inch in diameter.
12.29 c	A paper wasp nest is not surrounded by a paper-like covering. It is just a comb attached to a surface with a stalk. The stalk has repellent chemicals to prevent ants and other insects from attacking the young in the nest.
12.30 d	Lady beetles aggregate when preparing to find a place to spend the winter. They tend to move in large groups in the fall of the year and aggregate on the warm, sunny side of a building. They squeeze through small cracks and crevices around doors and windows. They also enter under clapboard siding.

13 Stored Product Pests

13.1 b, c and d	All types of food and non-food items derived from animal and plants which are storable for a long period of time are the primary target of stored product pests. However, wood infesting pests are not categorized in this group.

13.2 a	Apart from insects, a few species of mites (order Arachnida) are also classified as stored product pests.
13.3 a and c	Beetles are the largest with 87 species followed by moths 23 species, mites 12 species, fly 1 species and Psocids 2 or more species. This is based on Mallis and Ebeling.
13.4 a, b and d	Mites are microscopic and often remain un-noticed by untrained eyes until the population becomes large. They infest grains, flour and cheese among many stored items.
13.5 a-2, b-1, c-5, d-3, e-4	
13.6 b	
13.7 b	The species got its name from being a serious pest of stored medicinal herbs.
13.8 c	The female uses strong mandibles to chew a hole into a grain kernel after which she deposits a single egg within the hole, sealing it with secretions from her ovipositor.
13.9 c	Adults do not cause damage. The larvae are surface feeders and produce lots of webbing.
13.10 c	Both the larvae and adults bore irregularly shaped holes into whole, undamaged kernels and the larvae or immature stages may develop inside the grain.
13.11 a	The mandibles of pea and bean weevils may be elongated, but they do not have the long snouts characteristic of true weevils.
13.12 a	Most stored product moths are active fliers unlike clothes moths which avoid light and usually are seen only in closets.
13.13 a and c	Red flour beetles are dark brown in color with gradually clubbed antennae, with a three-segmented club. Confused flour beetles are reddish-brown with antennae gradually enlarged toward the tip, producing a four-segment club.
13.14 a-4, b-2, c-3, d-5, e-1	
13.15 d	Fabric pests are unusual in the insect world because they are among very few insects that produce an enzyme in their digestive system. This enzyme, keratinase, allows them to digest keratin, the protein in animal hair.
13.16 a and b	Depending on the type of food processed and type of stored ingredients, the risk of internal or external feeder can easily be identified.

13.17 b	It has been estimated that between one quarter and one third of the world grain crop is lost each year during storage. Much of this is due to insect attack.
13.18 a and d	Light traps do not attract all types of stored pests and pesticide application is only recommended when infestation exceeds a desired limit.
13.19 c	These insects produce a foul odor and taste in the food products they infest, which is caused by toxic quinone compounds produced by the pest.
13.20 b	Glue traps consist of a sticky glue layer mounted on a piece of cardboard that is folded into a tent-structure to protect the sticky surface. Most sticky traps contain no pesticides, and may be impregnated with aromas designed to be attractive to certain pests.
13.21 c	Phosphine is the most common fumigant suitable for grain fumigation. Its ease of handling and effectiveness have made it the most important fumigant.
13.22 a-1, b-3, c-1, d-1	
13.23 b and c	Combining a low level of phosphine with a moderate level of heat and high carbon dioxide has shown higher efficiency in controlling stored product pests. Both heat and carbon dioxide help increase the stress levels on insects.
13.24 a and b	
13.25 a	Phosphine is unreliable for psocid control. It may also explain the rapid resurgence of psocid infestation following phosphine fumigation, often observed due to elimination of natural predators and competition.
13.26 a, c and d	Sanitation is a primary method to prevent stored product infestations.
13.27 a	The removal of infested material will help the spread of the pest. Stored product pests are the slowest movers among pests so this action helps keep the infestation in check.
13.28 a, c and d	Insect light traps uses UV-A to attract insects. Their installation and servicing are critical to making them an effective tool.
13.29 a	Most stored product pests are attracted to light and are capable of varied levels of flight.
13.30 b	The key is to rapidly increase the temperature to 50°C and hold the temperature for 1–8 hours depending on the species or developmental stage of the pest.

14   Rodent

14.1 b	Due to its coloration and the fact that these rats can be found living around sewer systems and other similar places, the Norway rat is also called the brown rat or sewer rat.
14.2 c	Because this rat likes to live in high areas it is sometimes referred as the roof rat, but is also given other common names due to its habits and food preference.
14.3 b	Despite its name, this rat did not originate from Norway. It occurs in many different regions of the world and has been causing problems to humanity for a long time. The rat is originally from central Asia.
14.4 d	This rat is arboreal, living naturally on top of trees in its native region in South-east Asia. It arrived in the Americas with explorers from Europe in the late 15th century.
14.5 b	The Norway rat is relatively large and its tail is shorter than its body length.
14.6 d	When intact, the long tail in black rats gives them a sleek appearance compared to the more stout Norway rat.
14.7 c	Despite the fact that Norway rats are much more aggressive and stronger, the two species can coexist because the black rat is well adapted for arboreal living, preferring elevated locations on roof and tree tops, and being less dependent on food and other resources from humans.
14.8 a	Despite the normal litter containing 6–12 young, some litters can have larger numbers, although that is not very common. The roof rat has larger litter sizes in warmer areas, while the Norway rat will normally have larger litters in colder temperate areas of the globe.
14.9 b	Although under laboratory conditions, with proper nutrition and care, rats can live for more than two years, in the wild, lifespans are much reduced and only a small percentage of rats live to see their first birthday.
14.10 c	Rats become sexually mature after 3 months of age and females can give birth every 24–28 days, although in wild populations the intervals between litters is longer.
14.11 a	Rats are at their peak of activity at the time of day when the light is changing, either from dark to light or from light to dark, but they can shift their period of activity to adjust to human activity, competition with other animals, and the availability of food.

14.12 a	The Norway rat is originally from open grassy steps, where no trees are available; therefore this rat has a tendency to nest in ground nests. The roof rat originates from South-east Asia, where trees are abundant. This species has developed the ability to nest in higher places to fit its original habitat.
14.13 d	Besides nesting in trees, roofs and attics, roof rats may also nest in voids in various locations.
14.14 b	The Norway rat has a preference for cereal grains, and the roof rat prefers snails, nuts and fruits. However, rats will eat a diversity of animal and plant foods, and other materials, including processed human food and products they encounter in the environment in which they live.
14.15 c	Depending on their habitat, rats can forage for food and feed at any time of the day, but are usually most active at times of low natural light at dawn and dusk.
14.16 d	Rats need between 30 and 60 ml of water per day per adult under normal conditions, but the amount of water required varies considerably depending on temperatures, the rat activity level, and diet.
14.17 a	Rat populations in natural habitats in the wild usually have a migration pattern to follow the food supply, and other cycles that may change food and shelter availability.
14.18 d	Under indoor situations and optimum conditions, house mice can breed during all months of the year and build populations very fast due to a fast gestation period (18–21 days), litters that can have more than ten infants, and the possibility of up to ten litters per year for a female.
14.19 b	If resources become scarce, and there are life-threatening circumstances such as the presence of predators and intense competition from other mice, life span can be considerably shorter.
14.20 a	House mice will consume just about any type of food they come across, including plants seeds, cereals, insects and other small animals, carrion, and any food that is available in the environment.
14.21 b	The house mouse can acquire enough water from its food and from the production of metabolic water.
14.22 c	Rats and mice orient themselves more by touch and smell because their vision is relatively poor.
14.23 c	Being a larger species, Norway rat feces is larger (~20 mm long) with rounded ends, whereas the smaller roof rat has smaller (~13 mm long) feces with pointed ends.

14.24 a	Both species produce a large number of fecal pellets daily, depending on the availability of food, with the house mouse normally producing between 40–100 fecal pellets per day, and the Norway rat producing 40–50 pellets per day.
14.25 d	Grease marks are a typical way to determine the presence of rats or mice in a location. These marks will appear along the walls or on other locations along the runways that are constantly used by mice and rats.
14.26 d	Gnawing damage can occur not only in wood but also on aluminum, plastic, lead pipes and even in mortar and other materials, and the size of the gnawing marks can serve as identification of the type of rodent present in the infested location.
14.27 b	Places infested with rodents normally carry a characteristic odor that can be recognized by experienced technicians, due to the accumulation of urine and fecal spots throughout the area.
14.28 b	Rodent control must combine several sanitation procedures to deny food to the rodent population, rodent proofing of the facilities to prevent rodent entry, and chemical baiting and trapping to eliminate any rodents that are able to gain entry into the facilities.
14.29 c	First generation baits must be available for a long period since the animals have to feed on them over 3–10 days so that the bait can be efficient in eliminating the rodent population.
14.30 a	Non-anticoagulants rodent baits either function as a nerve poison, or deplete the calcium from bones. Their advantage is that there is less of a chance of secondary poisoning in relation to anticoagulant baits.

15   Birds and Bats

15.1 a-1 and 4, b-2 and 3	The commonest pest birds in structures are pigeons and house sparrows. They differ in feeding and nesting habits and are often controlled by following different strategies.
15.2 c and d	The most important health risk pest birds cause is the spread of diseases carried in their droppings.
15.3 a	
15.4 e	
15.5 c	A survey of the infestation is a must before deciding on the management strategy. A survey reveals bird species, number, nesting, feeding habits and access points to the structure.
15.6 a, c and d	Bird surveys should be conducted 2–3 times in a single day to have a complete view of bird activity and numbers.

15.7 a	
15.8 a-2, b-3, c-5, d-1, e-4	
15.9 a, b, and c	Mist nets are made of nylon and used to capture wild birds. They are not used as an exclusion tool.
15.10 a, b and d	The purpose of using bait is to kill a certain percentage of pest birds and make the rest wary, encouraging them to relocate.
15.11 b, c and d	Bait shyness is common when birds witness convulsions and distressed flock members. This makes them frightened and they avoid baits for a duration.
15.12 b and c	Trapping can reduce isolated pest birds and can be used as an effective supplemental method.
15.13 c and d	A higher amount of Avitrol allows faster movement of the flock to another spot. However, this may also increase the visibility of dead birds, which should be avoided.
15.14. a, b,and c	Exclusion is the best method to keep pest birds away from a property and building design plays a vital role in ensuring it.
15.15. a, b and d	Only a few species of bats colonize structures as roosting sites. These bats mostly live in colonies or groups.
15.16 e	
15.17 b	Chiropterophily means pollination which occurs through bats and the flower is known as the chiropterophilous flower. Bat pollination is commonly found in tropical and desert areas.
15.18 a	Bats produce sound waves at frequencies above human hearing, called ultrasound. The sound waves emitted by bats bounce off objects in their environment. Then, the sounds return to the bats' ears, which are finely tuned to recognize their own unique calls
15.19 b	Very few items truly repel bats. Naphthalene is the only product registered as a bat repellent. However, the product works best in confined spaces.
15.20 c	Both the diseases are associated with bats but the rate of transmission is not considered high.
15.21 a and b	The best solution it to capture and release the bat if trapped inside. A gloved hand may be used too but care needs to be taken when handling the animal.

16   Pesticides and Formulations

16.1 b	An insecticide is any chemical that kills insects. The "icide" part of the name indicates that it kills and the preceding part of the name tells what kind of organism the chemical kills. So, a pulicide kill fleas, a rodenticide kills rodents, and a herbicide kills plants.
16.2 a	The general structure of an organophosphate is the element phosphorus combined with oxygen to form the base of the molecule.
16.3 b	A common carbamate is carbaryl. Carbaryl has been used for years in turf and ornamental applications. It has been a common product used in garden dusts.
16.4 c	Nerve synapses connect one nerve cell to another. The nerve impulse crosses the synapse on acetylcholine (ACh). Acetylcholinesterase (AchE) normally breaks ACh into acetate and choline to stop the repeated firing of nerves. The inhibition of AChE by organophosphates and carbamates results in repeated nerve firing, twitching, and ultimately the death of insects.
16.5 c	Pyrethroid insecticides are derived from the chemistry of the chrysanthemum flower. They typically have a cyclopropyl group in the molecule. The cyclopropyl ring is a 3-carbon ring that looks like a triangle when the chemical structure is pictured.
16.6 a	Pyrethroids affect the axons of nerve cells causing repeated nerve firing and excitation in affected insects. Specifically, pyrethroids prevent the voltage-gated sodium channels along the axon from closing. By keeping the channel open, the voltage remains charged which causes super-excitation of the nerve.
16.7 d	Neonicotinoids have a structure similar to nicotine. The active part of the molecule has nitrogen in a nitroguanidine and cyanoamidine configuration. Nitrogen is an important element in both neonicotinoids like imidacloprid and the parent molecule, nicotine. Neonicotinoid insecticides are usually soluble in water. As a result, they are systemic in plants and effectively control aphids and other sucking insects. They have also been widely used in topical flea control products.
16.8 c	Bifenthrin is a pyrethroid insecticide. The common neonicotinoid insecticides for the pest control industry are imidacloprid, acetamiprid, and thiamethoxam. Also, dinotefuran is a neonicotinoid that is highly effective for urban pests.
16.9 d	Thiamethoxam is not a pyrethroid insecticide. You should notice that typically the pyrethroid insecticides have names that end in "thrin".

16.10 a	
16.11 b	
16.12 d	
16.13 c	It is necessary to switch away from a pyrethroid, because bifenthrin is a pyrethroid that has been used for years. The non-pyrethroid in the list is imidacloprid. It is in a different chemical group and has a completely different mode of action from the pyrethroids.
16.14 a	Boric acid is a common active ingredient in baits and dusts. It is usually ingested or groomed off insects. Once ingested, it kills cells in the gut causing eventual death of the pest insect.
16.15 b	Vacuuming pests, like cockroaches, is a good way to remove about 80% of a severe infestation without selecting the population for resistance. Other methods of physical control could be the use of heat or cold to kill pests. Ultrasonic devices have never been proved to be effective for pest control. The use of natural enemies is a form of biological control and is not a form of physical control. Fumigation is a form of chemical control.
16.16 a	Concrete and concrete blocks have a high pH (>10 pH). Insecticides readily break down at high pH and leave virtually no residue. Emulsifiable concentrates diluted in water to form an emulsion would soak into the concrete and break down due to the high pH. ECs would be a poor choice to apply to concrete. Wettable powders, suspension concentrates, and granules could be applied to the surface and not be readily influenced by the pH of the substrate.
16.17 d	Total release aerosols have been shown to be ineffective for bed bug control. They also have been shown to be ineffective for some cockroach populations. These aerosol bombs usually have pyrethrins or pyrethroids as their active ingredients.
16.18 d	Insecticide dusts applied under appliances with fans can be blown onto food or food preparation surfaces by the air movement. It is important that dusts be applied into voids or cracks and crevices where pests harbor.
16.19 c	Most total release aerosol bombs have an explosive propellent like butane. When the concentration of propellent in a room is excessive, a spark from an appliance starting or a pilot light on a stove can set off a large explosion that can blow out the windows and doors of a building.

16.20 e	All the items listed are needed for a thorough inspection. Pests hide in cracks and crevices and are difficult to find unless proper inspection equipment is used.
16.21 a	Any introduction of yourself to a customer goes best with a smile and confident eye contact. That sets the stage for a professional greeting and appropriate handshake. Later, you can also use some sincere compliments to show your interest in their situation.
16.22 b	Juvenile hormone analogs like methoprene and pyriproxyfen are used to sterilize cockroaches and kill fleas and mosquitoes as they pupate. They are not very good at controlling occasional invaders because they are rather slow acting and would not prevent these pests before they gain access to the structure. They are not used for either bed bug or termite control.
16.23 a	Chitin synthesis inhibitors are used in most termite baits. They are really slow working and kill termites as they try to molt from one stage to the next.
16.24 c	Insecticide resistance develops over generations of pests due to exposure to the same active ingredients or insecticide groups over time. So, rotation of insecticide groups is essential to preserve the use of an insecticide long term.
16.25 a	The diamides like chlorantraniliprole, are used for control of termites in urban pest control. The diamides affect the ryanodine receptors at the muscle synapse causing muscle paralysis.
16.26 c	The best way to avoid risk in applying insecticides is to apply products when pests are present and to concentrate on areas where they occur.
16.27 b	Due to Covid, many people are used to respirator terminology. An N95 particle respirator will prevent inhalation of aerosol particles of insecticide. N95 refers to the type of particle and the percentage of particles removed from the air. The "N" means that is will work on non-oil based insecticides, and 95 means that 95% of those type of particles will be removed.
16.28 d	Start a pest inspection at a location and move systematically around the location. Take notes on where pests occur and where they do not.
16.29 b	The first step in handling an insecticide spill is to read the label. That information will tell you the correct PPE to use for the product.
16.30 d	Pesticides should always be stored in a secure location. You should not put them in the bed of a truck where they could be stolen or accessed by children. They should always be stored in the lock box. Labels and SDSs for the products can be stored in the passenger compartment so they are handy for reference or emergencies.

17 Handling Pesticides

17.1 c	The LD-50 is the amount of pesticide or lethal dose that will kill 50% of the tested population in a standard time interval (usually 24 h). The higher the LD-50, the less toxicity. The lower the LD-50, the higher the toxicity of a product.
17.2 a	Toxicity category I is a product with an LD-50 of less than 50. A pesticide with that rating is a category I pesticide and is highly toxic. It also could be severely irritating. Toxicity category I pesticides are classified as restricted for use by professionally trained and certified applicators.
17.3 b	Toxicity is a function of dose multiplied by time. This means that if you lower the dose, it takes longer for mortality to occur. Conversely, if you increase the dose, adverse effects occur sooner. That is why it is important to use all the safety equipment that is recommended on the label.
17.4 c	The NOEL stands for the no observable effect level. If a dose is sufficiently low, the time period for producing an adverse reaction may exceed the life span of the species.
17.5 d	Low doses of chemicals can induce cells to produce tumors. These tumors may be benign or malignant. Neither of these traits is considered beneficial. So a chemical may be considered dangerous due to its oncogenetic properties.
17.6 b	Children usually encounter pesticides that are stored in the house and then drink or eat them. This oral exposure is the most common way that children encounter toxic pesticide doses.
17.7 d	Dermal exposure is one of the most common routes of entry for pesticide applicators. Gloves are used to protect hands and arms from splashes while mixing and loading pesticides into application equipment.
17.8 c	Pesticides are often applied to the air to kill flying insects and also exposed crawling insects. These pesticides are airborne and easily inhaled. It is important to use a respirator to prevent exposure by inhalation of the insecticide.
17.9 a	It is quite common to measure liquid insecticides into measuring cups by holding the container up to eye level. Usually this exposure is prevented by using safety glasses, goggles, or face shields. Ocular exposure can cause eye damage and is an important route of pesticide exposure.

17.10 b	One of the greatest risks for applicators is eating or smoking without washing hands. Pesticide mixing or application can result in small quantities being deposited on hands. These pesticides should be washed off before eating or smoking. There is very little risk of poisoning when hands are washed after pesticide use.
17.11 c	The top risk for ranked everyday activities was smoking. You would think that pesticides would be ranked as highly dangerous. But of the top 30 daily activities, pesticides were ranked number 28.
17.12 a	Relative toxicity of insecticides is expressed as mg of active ingredient per kg of body weight. An LD-50 of less than 50 is considered highly toxic. The toxin produced by the botulism bacteria (used in Botox) has an LD-50 of 1–3 nanograms per kg of body weight. That is more than a thousand times more toxic than any insecticide.
17.13 d	EPA is the federal agency that regulates insecticides in the US. EPA can register a product after it has been proven effective and would not adversely affect people or the environment.
17.14 b	OSHA promulgates regulations to inform workers about hazards of the job. In the case of pest control, the regulations inform workers of chemical hazards in the workplace and protective measures to prevent injury. Even though labels provide safety information, OSHA develops standards for providing safety information to workers.
17.15 c	The label provides information on how to use a product effectively and safely. It has various sections to provide necessary information about where to use the product, how to protect yourself, how to use the product, and how to store and dispose of the product.
17.16 a	The safety data sheet (SDS) provides information on what to do in the case of an emergency. There will be information on how to deal with a pesticide spill. The SDS will give specific hazards and protective measures in the case of an accident, like a spill.
17.17 c	There are pictograms on the SDS to indicate the type of hazard associated with the product. One product may have several pictograms if the product has several types of hazards. These pictograms indicate explosion, fire, irritation, etc. You should get to know and recognize these pictograms.
17.18 c	There has been a lot of research on how to launder pesticide contaminated clothing. The best advice is to launder pest control service clothing on a regular basis (daily if possible) with hot water (140°F) and rinsed with either hot or cold water. A heavy-duty detergent is recommended.

17.19 d	The pesticide storage facility should be labeled with DANGER: PESTICIDE STORAGE AREA, if the label for the most dangerous product in storage says DANGER. If the most dangerous product says WARNING, then the area should be labeled WARNING: PESTICIDE STORAGE AREA. So the posting should be in accordance with the label of the most hazardous product.
17.20 b	Pesticide spills should never be allowed to get into sewers or septic tanks. A pesticide storage area should not have any drains and the floor should be designed to contain any spills.
17.21 b	Work shoes would be an appropriate protection for feet, but athletic shoes would not give protection from spilled chemical. A golf shirt is usually a short-sleeved shirt and would not meet the label requirements. Also, disposable sleeves are not an adequate replacement for the required long-sleeved shirt. Most regulators have decided that disposable sleeves are not a long-sleeved shirt.
17.22 b	The new EPA labels rate materials for gloves according to pesticide solvent categories. A table is available online or in print. According to the chart, polyvinyl chloride is an appropriate material for an "E" category pesticide product.
17.23 d	There are many ways to protect eyes from pesticides. In some cases, the label may not require anything more than protective glasses. However, some products are very irritating to the eyes. These products have a statement like "Causes moderate eye irritation" or "Causes irreversible eye damage". Always read the personal protective equipment statement for each product that is used.
17.24 d	NIOSH rates respirators for their efficiency in protecting workers from aerosols. For Covid, the N95 respirator was recommended because N stands for non-oil aerosols, like particles in a sneeze. However, many aerosol insecticides are diluted in oil and require an oil resistant filter.
17.25 d	Particle filters are labeled as magenta in color. These filters will not protect an applicator from pesticide vapors.
17.26 a	Vapor filters for respirators are color coded to determine the type of vapor it protects against. Most insecticides are organic vapor filters. Of course, magenta does not protect against vapors, just particles.
17.27 b	The best color combination of filters for an insecticide aerosol treatment would protect against both particles (aerosol droplets) and insecticide vapor. So the correct answer would be magenta to protect against particles and black to protect against organic vapors. The correct respirator is recommended on the product label.

17.28 a	The label is the law. The personal protective equipment required to use the product is listed on the label. So any time that the label is not followed, even to protect the applicator, is a violation of the label. So the law would be broken if the correct PPE is not utilized.
17.29 d	The federal insecticide, fungicide, and rodenticide act, also known as FIFRA, regulates pesticides in the US. FIFRA regulates the sale, distribution, and use of pesticides in the US. Most pest control falls into the pesticide use part of the law. The law provides for certification of pesticide applicators and requires users to follow label directions. EPA regulates pesticides under the provisions of FIFRA
17.30 b	In the USA, the states are responsible for enforcement of the Federal Insecticide, Fungicide and Rodenticide Act.

18 Integrated Pest Management

18.1 b	Integrated Pest Management involves the use of several techniques beside chemical pesticides, in order to achieve safe and efficient control of pests.
18.2 b and c	Integrated Vector Management (IVM) is the main method for tackling many of the world's most burdensome infectious diseases, such as malaria, dengue and other neglected tropical diseases (NTDs) by targeting mosquitoes, flies, ticks, bugs and other vectors that transmit pathogens.
18.3 c	IPM is not a single pest control method but integrates a number of pest management methods including evaluations, decision making and controls.
18.4 e	IPM integrates human behavior as an important criterion in designing programs.
18.5 a	Builders, architects and designers rarely take pest matters into consideration when designing a structure.
18.6 a, b and c	Pests are attracted to a structure due to a number of physical features as well as to find shelter. These, if modified or avoided, will help.
18.7 a	Different kinds of light contain different amounts of ultraviolet (UV) light. UV is attractive to the majority of night flying insects. With a knowledge of light types, one can strategically reduce the number of insects.
18.8 a	Studies have proven that customized IPM can be successful and cost-effective compared to industry practiced methods.

18.9 a, b and c	In the late 1960s and early 1970s, awareness of the harm that chemicals like DDT had created prompted the start of the environmental movement, including IPM, around the world.
18.10 b, c and d	The book exposed the hazards of the pesticide DDT, eloquently questioned human dependency on chemicals, and helped set the stage for the environmental movement around the world.
18.11 b and c	Meta-analysis revealed a positive association between exposure to residential pesticides in pregnancy and childhood leukemia.
18.12 a, b and c	Proper design, landscape and using stainless steel mesh to exclude termites, putting caps on roof tiles to reduce bird and rat infestations, and sloping windowsills to discourage pigeons, can all significantly reduce infestation.
18.13 b, c and d	Record keeping is a must in IPM programs, which take note of every individual activity.
18.14 b	Communication between property owners and the IPM managers reinforces actions the homeowners must take in order to prevent pest problems, and provides a line of communication so the homeowners can provide feedback on the effectiveness of the pest management program.
18.15 e	Humans face risk from accumulative exposure to pesticide than individual ones, which is often difficult to ascertain.
18.16 e	All the above reasons, including an increase in pest prevalence, is making consumers choose pesticides as a solution.
18.17 b and c	Repeated use of the same class of pesticides to control a pest can cause undesirable changes in the gene pool of a pest leading to pesticide resistance, making the product ineffective.
18.18 a and d	Food-based baits are also very important in the control of building-infesting ants.
18.19 c and d	Food-based insecticide baits are species specific and best against cryptic pests. Due to a process of horizontal transfer, the insecticide is quickly spread among group or colony members.
18.20 a	The technician inspecting the structure is the best person who can determine the correct status of the pest under the prevailing conditions.
18.21 b	In a structure, pests are concentrated in places where their most needed resources are available. Kitchens have the greatest number of German cockroaches, whereas bed bugs will be in the highest number in bedrooms.
18.22 a	Conventional pest management generally means the spraying of pesticide to irradicate a pest.

18.23 a, b and c	Pest knowledge is critical in correct judgement and sound decision making which are the basis for IPM.
18.24 c	Properly implemented IPM programs are successful in both the fields.
18.25 d, c, b, a	The first step is designed so the pest control operator is very familiar with the present pest situation, before proceeding with measures that will minimize the pest problems starting with non-pesticidal methods, and using pesticides only as a last resort.
18.26 d	IPM needs customization and specific details to be successful, not a pre-programmed work scope.
18.27 a-3, b-2, c-4, d-1	
18.28 b and c	Psychological resistance to change and resistance in learning new technologies, are among reasons for poor adoption of IPM.
18.29 b	Often surface spray and baiting will not go hand in hand so baits can be used along with other methods to provide a satisfactory level of control.

CABI – who we are and what we do

This book is published by **CABI**, an international not-for-profit organisation that improves people's lives worldwide by providing information and applying scientific expertise to solve problems in agriculture and the environment.

CABI is also a global publisher producing key scientific publications, including world renowned databases, as well as compendia, books, ebooks and full text electronic resources. We publish content in a wide range of subject areas including: agriculture and crop science / animal and veterinary sciences / ecology and conservation / environmental science / horticulture and plant sciences / human health, food science and nutrition / international development / leisure and tourism.

The profits from CABI's publishing activities enable us to work with farming communities around the world, supporting them as they battle with poor soil, invasive species and pests and diseases, to improve their livelihoods and help provide food for an ever growing population.

CABI is an international intergovernmental organisation, and we gratefully acknowledge the core financial support from our member countries (and lead agencies) including:

Discover more

To read more about CABI's work, please visit: **www.cabi.org**

Browse our books at: **www.cabi.org/bookshop**,
or explore our online products at: **www.cabi.org/publishing-products**

Interested in writing for CABI? Find our author guidelines here:
www.cabi.org/publishing-products/information-for-authors/